Park Life

A Year in the Wildlife of an Urban Park

"It gives me a sense of enormous well-being"
('Parklife' by Blur).

Rick Thompson

Grosvenor House
Publishing Limited

This book is published by
Grosvenor House Publishing Ltd
Link House
140 The Broadway, Tolworth, Surrey, KT6 7HT.
www.grosvenorhousepublishing.co.uk

A CIP record for this book
is available from the British Library

ISBN 978-1-83975-173-8

Contents

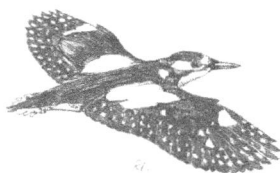

Preface

For a quarter of a century I had lived in a cottage overlooking the fields and woods of Warwickshire. As a lifelong birdwatcher and naturalist, I loved walking the country lanes and observing the seasons unfold. So when my wife and I moved into the centre of the county town in order to be in a bustling community, I feared I would be isolated from nature. It was a revelation to discover that the 100-acre park beside the River Avon in the centre of Warwick teems with wildlife, some of it very unexpected!

This is my diary of one year in the life of the park with tips on how to identify birds, insects and plants, the intriguing folklore associated with much of our wildlife, and reflections on why regular contact with the natural word is so important for our physical and mental wellbeing.

I hope readers will enjoy accompanying me on my walks in the park, learning about the amazing creatures that can be found in an urban setting, enjoying the changing seasonal moods along the riverbank, discovering rarities, and seeing how climate change is already having an impact on the species to be found in a typical town park in the centre of England.

I could not know that the year of my wildlife diary was to be followed by the year of the virus, when everything changed. But the lockdown was to show vividly how important it is for us all to get out into green spaces and connect with nature. The silence and the clean air soon made everyone aware of the birdsong, so often ignored or obscured by the noises of the city. It should not take a crisis to awaken us to the natural wonders on our doorstep. Even in the centre of a town, nature flourishes all year, every year, in its extraordinary cycle. We just have to look and listen.

Rick Thompson

ROUGH FIELDS

KINGFISHER POOLS

HOUSING ESTATE

FOOTBRIDGE

THE COPSE

MYTON FIELDS

MYTON ROAD

BEECH AVENUE

RIVER AVON

LEISURE CENTRE

POOHSTICKS BRIDGE

PLAYGROUND

FUNFAIR

ST. JOHN'S BROOK

BANBURY ROAD

ST. NICHOLAS CHURCH

GATE

A 425

WARWICK CASTLE

THE COLLEGIATE CHURCH OF ST. MARY

THE PARK

E
N — S
W

Introduction

From Country to Town

"I wonder what life is like there?" thought
Country Mouse.
(Town Mouse; Country Mouse. BBC Jackanory)

Lapwing

Making the Move

I stood at a gateway watching the lapwing swooping and tumbling over the beige and sepia field as it called out its liquid 'peewit' country name, and wondered if it would breed successfully this year. But I would never find out. After a quarter of a century living in the cottage overlooking the Warwickshire fields, we were moving. The van was due in half an hour.

This was a final farewell to the country lane I knew so well. Part of the Heart of England Way, it was probably an old drovers' lane, with wide borders and high blackthorn hedges, and ribbons of hedge-row running away across the fields; or as Wordsworth described, "*hardly hedge-rows, little lines of sportive wood run wild*". I scanned the uncultivated field with its rough grass and head-high thistles where, year after year, the lapwings had failed to fledge any young; too many buzzards and crows around probably. This was also where the ghostly barn owl, white against the dusk, had silently quartered at twilight, coming very close if you stood stock still. There was the gate to the barley field where a grasshopper warbler had reeled endlessly one spring, while the tangled hedgerows along the lane had been alive with the scolding of whitethroats and blackcaps. And this patch of scrub was where I had startled the elusive woodcock. Or rather it had startled me, suddenly rocketing away from under my feet and darting over the hedge.

As our boys had been growing up they too had loved the rural setting, running across the fields with their friends, making dens in the woods and skateboarding down the quiet lane. When they had left home to make their way in the world, my wife had started to drop hints that we too should move on. "There's nothing here. We have to get in the car to get a pint of milk. We should move into a town where everything is close by

2

and where we can get involved in a real community." She was right.

Some people when they reach retirement age talk of moving to a cottage in the country, ideally with roses climbing round the door, or to a bungalow by the sea. I think this is often a mistake. As one gets older, surely it's a good idea to live where there's a variety of shops within easy walking range, not to mention a doctors' surgery, a dentist, perhaps a post office, ideally some outlets for local food such as a butcher's and a greengrocer's, and certainly some decent pubs and restaurants.

We decided to move to Warwick just a few miles away. It fitted the bill nicely. The county town of Warwickshire is compact – it used to be a medieval walled town – and it is noted for its independent shops, particularly along Smith Street that survived the Great Fire of Warwick in 1694. It has a Saturday market with local produce, and countless pubs and restaurants.

It's certainly a thriving community with annual events that bring in thousands of visitors, including a major Folk Festival, Food Festivals, a Beer Festival and the Warwick Words History Festival. The big tourist attraction is Warwick Castle, one of the largest and best preserved Norman fortresses in Europe. More than 800,000 visitors pass through its gates each year.

We were fortunate to find a house in the very centre of town, next to the Collegiate Church of St. Mary. Its tall and elegant tower at the top of the hill can be seen from miles around. Welcome to historic Warwick, and urban living.

Discovering the Park

So where would my morning walk take me now? Would choking traffic fumes and noisy buses replace the breeze rippling the wheat across the wide open fields and the cries of the lapwings? What a pleasant surprise was in store. I soon found that Warwick is blessed with two large parks, both within a few minutes' walk from our front door. On rising ground to the north-west of St. Mary's is Priory Park, named after the 12th Century Dominican Priory of St. Sepulchre that occupied the land before Henry VIII dissolved it. Later the stones of the old priory were transported to Richmond, Virginia, to be re-erected as a stately home for a rich businessman. Now Priory Park has undulating areas of open grass, some deciduous woodland and large patches of tangled scrub. It is full of rabbits and squirrels, much to the delight of the dogs walked there each day.

The other park is larger and is the main open space in the centre of Warwick. Standing next to St. Nicholas' Church, which was historically the place of worship of the Earls of Warwick in the nearby castle, St. Nicholas Park turned out to be an unexpected treasure. St. Nick's, as it is known by the locals, is nearly 100 acres. Originally a meadow by the River Avon, it was purchased by the council and laid out as a park in the 1930s. After the Second World War, there was a strong movement to give the post-war kids every advantage, including the opportunity for healthy pursuits; so urban parks with playgrounds and sports fields were developed everywhere. St. Nick's in Warwick was typical, with a play area, a paddling pool and outdoor swimming pool, tennis courts, and football fields. Later the swimming pool was replaced by a large leisure centre with an indoor pool, gyms and sports halls. There's even a small funfair with rides for small children.

In the post-war years, the planting of trees was financed by relatives of local servicemen who had died in both World Wars. More than seventy years on, there is a magnificent avenue of mature copper beeches, and lines of sponsored weeping willows beside the river. The commemorative plaques are now embraced and sometimes overwhelmed by tree roots, but some of the names of those who gave their lives can still be deciphered.

How could wildlife flourish in such an environment geared to human fitness and leisure? I soon discovered that St. Nicholas Park, Warwick, had much more to offer than tennis courts and swings. If you walk away from the play area and along the riverside path, everything changes. There is a surprising variety of natural habitats. On the river bank there is a copse of birch, silver birch, oak, sycamore, and yew. Alder and willow lean over the water. There are clumps of hawthorn and briar overshadowed by tall stands of ash and oak.

Over a footbridge to the south side of the river are The Kingfisher Pools, with wooden platforms for the anglers, and large stretches of water fringed by patches of reed bed and thicket. A finger of land bordered by bulrushes stretches into the centre of the main pool. Behind a ribbon of tall trees at the end of the park is a neglected patch of rough ground, two fields seemingly untouched for years, rampant with thistles and low scrub, and at the fringes great mounds of brambles. Open grassland, mixed woodland, reed bed and sedge – this is promising for wildlife. And then there is the river!

The Avon

The River Avon is sometimes known as the Warwickshire Avon to distinguish it from other rivers of the same name. The source is at Naseby in Northamptonshire and it flows south-west to join the River Severn at Tewkesbury. In Warwick it passes over a natural weir under the walls of the castle and heads towards Stratford. War-wick means the settlement by the weir, and dates from the construction of a burgh or fort here by Alfred the Great's daughter, Aethelflaed, the mightily impressive Lady of the Mercians. She constructed a chain of forts along the frontline of the long-running war with the Danes to the north-east. The Anglo-Saxon Chronicle records that after building forts in Stafford and Tamworth, in the year 914 she had one constructed at Eadesbyrig (Eddisbury in Cheshire), *"... and afterwards, in the same year, late in harvest, that at Warwick"*. So Aethelflaed is regarded as the founder of Warwick more than eleven hundred years ago.

There are five River Avons in England, three in Scotland and one in Wales. Why? Well the name "Avon" comes from the Welsh word afon meaning 'river'. You can imagine the invading Romans saying to the local Celts, "What do you call that?" (pointing at the river). And they would blurt out, "Avon". "Right, lads", says the Centurion, "It's called the River Avon". In reality it means 'River River'.

Here in St. Nicholas Park, the River River, (The Warwickshire Avon), is a constant source of interest. There was no chuckling river where we used to live. Just the motionless and rather muddy waters of the Grand Union Canal. The river running through the park is much more interesting, with distinct moods at different times of year, as well as its own wildlife.

Exercise and Therapy

I soon took to walking round the park every morning. It became my 'patch'. I wondered how many bird species I would be able to identify here in the centre of a town? Would there be interesting insects such as butterflies and dragonflies in the summer? There was clearly a wide variety of plants alongside the river and the fishing pools. Could I identify them? I began to make notes of what I had seen and heard in the park on my daily walks and found that watching the seasons unfold was just as rewarding as when I had been walking along that country lane each day. In fact, more so. Nature was on everyone's doorstep in all its intriguing splendour. All the town-dwellers had to do was to look, and listen.

Various studies have consistently shown that regular contact with nature is good for you. The researchers call it 'exposure to greenspace'. In 2019 a team from the University of East Anglia assembled data from 140 studies in 20 countries to see if contact with nature really does boost your health. The researchers concluded that, "Spending time in natural green spaces is associated with diverse and significant health benefits. It reduces your risk of type II diabetes, cardiovascular disease, premature death, preterm birth, stress, and high blood pressure". Gosh! The research team suggested that doctors should advise their patients to spend more time in greenspace, rather than prescribing pills. "We hope this research will inspire people to get outside more and feel the health benefits themselves. Hopefully our results will encourage policymakers and town planners to invest in parks and greenspaces".

Relieving stress is an interesting benefit on that list. The NHS now prescribes 65 million items of anti-depressants each year, twice as many as 10 years before. Nearly 12 million working days are lost annually because of stress, anxiety or depression. It doesn't take an academic to tell you that the countryside and green spaces reduce the stress of modern life. Surely everyone feels an uplift of the spirits on a walk in a natural landscape, even in a large urban park.

Other research is rather startling about the way the growth of urban living has disconnected many people from the natural world. A 2019 survey of a thousand children aged 5 to 16 came up with results that I found rather depressing. More than half could not identify a stinging nettle and only 18% could identify an oak leaf. Two out of three could not identify a kingfisher and astonishingly 23% failed to identify a robin. The introduction of a GSCE in Natural History is long overdue. Whatever happened to the nature-table in the classroom and taking the kids out for pond-dipping or bird identification? I am afraid that years of cuts to school budgets, combined with an emphasis on testing and performance targets for traditional subjects, have left little time for city children to experience nature.

Walking the Walk

For older people, the daily walk in the park has significant other benefits, as well as cheering you up. I am not a fan of jogging when you reach a certain age. I know too many joggers with knee or ankle problems caused by the jolting. Innumerable experts advocate brisk walking as the best daily exercise. A study by the University of Warwick showed that those who walk more and sit less have a lower BMI, (Body Mass Index), which is an indicator of obesity. Regular walkers have a 12% lower risk of type-2 diabetes. Walking is linked to a 7% reduced risk of high blood pressure and even high cholesterol.

There's even more. A clinical trial in Japan concluded that after 3 months of daily walking, older people had significantly better memory and ability to concentrate. It seems that walking – especially in a natural environment – stimulates the production of neurotransmitters in the brain, such as endorphins, helping to improve your mental state. And a research paper published in 2014 concluded that walking for roughly 3 hours a week was associated with an 11% reduced risk of premature death!

So the daily walk in the park is going to keep me alive longer! And the contact with nature significantly reduces the stress of modern life. I guess I knew that without having to read the research papers. Closely observing the wildlife through the year is an uplifting pleasure.

I was to discover that with the changing seasons in my local park, there would be plenty of inspiring experiences, a fabulous range of species, and some big surprises. This is my diary of one year in a park in the very centre of England.

January

*'If winter comes can spring be
far behind?' (Shelley).*

Snowdrops

Thin Ice

After a very mild Christmas and New Year we have a brief cold spell and the fishing pools in the park are frozen with thin sheets of glass. A biting easterly breeze is hissing through the reeds. A raft of black-headed gulls is standing on the ice, looking on bemused as a pair of mallard land next to them, skidding and skittering on their tails as in a Walt Disney cartoon. The heron that appears regularly along this reach of the river stands disconsolately in the shallows, fluffed up against the cold. Why don't birds get chilblains? Or frostbite? Their unfeathered feet must be frozen!

It seems there's a good reason why even the coldest of winters aren't much of a problem for most birds. Their feet are ingeniously designed so that they're already cold to begin with, thanks to a network of arteries called *rete mirabile* or 'wonderful net'. A bird's heart is wired to its feet in such a way that by the time the tiny amount of blood gets down there, it's already cooled. A heat exchange system ensures warm blood stays close to the bird's heart, while the cool stuff dribbles down to its toes. The bird feels very little down there, and, most importantly, doesn't experience much heat loss. There are no muscles at all on a bird's lower legs and feet, only thin tendons. That means they hardly need more than a pitter-patter of blood to stop the blood vessels freezing. So now I know why they can stand on the ice and not call for thermal socks. They simply don't feel the cold.

The ice on the pool has some intriguing patterns, like pieces of modern art, with circular patterns, and crazed areas that look like stained-glass windows. They remind me of the windows in the celebrated Beauchamp Chapel in the nearby Collegiate Church of St. Mary that dominated the skyline with its tall tower. The beautiful windows were smashed by the

Parliamentarians in the English Civil War, but their pikes were not long enough to reach the top sections, which remain to this day as classic examples of ecclesiastical coloured glass. But putting back together the lower sections from the shards left behind must have been a difficult task for the canons of Warwick. They made a decent fist of it, though one portrait of a saint has two left hands.

The cold snap doesn't last long and the ice on the pools melts away. Westerly airstreams bring a canopy of cloud, flat and steel-grey. It's very mild and the birds all wake up. Sounds of spring are everywhere. The great tits are 'belling' their ding-dong spring song. They are supposed to say 'Teacher teacher', but I reckon they say lots of different things. Great tits can have a triple note or a single 'clink … clink'. If you have these birds in your garden, in the spring you may soon be able to identify individuals. Every calling male sounds a little different. So the great tit can be a bit tricky to identify, but the metallic 'belling' sound is always there. The smaller blue tits seem to be everywhere, chasing each other through the bushes with their constant high-pitched trills. If they were in a Carry On film, 'Carry On Tits' perhaps? the blue tit's giggle would be Barbara Windsor and the great tit's deeper suggestive snicker would be Sid James.

The little coal tits are also very vocal now, with a different higher-pitched call that sounds like a squeaky toy being squeezed. 'See-saw See-saw' – a remarkably loud voice for a bird that is even smaller than a blue tit. I love the coal tit with its peachy colouring and badger-striped head. Harassed by all the other birds on your garden feeder, this little artful dodger will dive in first when the others have been spooked into the bushes, or nip in and out the moment the other birds allow an opening.

Early Spring

In the park, signs of an early Spring are now everywhere. Already there are carpets of snowdrops nodding under the trees, and the crocuses planted by the District Council in great patches in the grassy areas are showing by mid-month. The Greek name for the snowdrop family is *galanthus* meaning 'the milk flower'. 'Galanthophiles' are snowdrop-lovers. I guess that must mean we are all galanthophiles. Who doesn't love this pretty early spring flower? Surprisingly the snowdrop contains a substance, galantamine, that is used to treat the symptoms of Alzheimer's Disease.

The mallards have paired up and are performing head-bobbing rituals which clearly say, "Alright? How about it?" … "Yes, I'm up for a bit of fun, but you'll have to wait until I'm ready". It's great to have such a beautiful bird as one of our commonest and easily seen wildfowl. Mallards are always worth a close look especially in the winter and early spring when their feathers are spanking-fresh. The drake has extraordinarily beautiful plumage in the breeding season with iridescent head-feathers changing from green to purple as they catch the light, and curled quiff feathers just above the tail.

Around the Kingfisher Pools the song thrushes are in full voice. It's a calm and mild morning and I count seven singing males competing with each other for territory and trying to attract females. From high perches in the trees, they each find an endless variety of phrases to repeat. It is a joy to pause awhile and listen carefully. They seem to be bursting with life. Gerard Manley Hopkins described the thrush's song rather effectively:

"..and thrush through the echoing timber does so rinse and wring the ear, it strikes like lightnings to hear him sing".

Too right, GMH. It does rinse the ear. Listening to thrushes in full-throated song in January is delightful and one can't help feeling optimistic.

The song thrush is still a red-listed or endangered bird after a steep fall in their numbers. According to the RSPB they declined by more than 50 per cent between 1970 and 1995. This was partly because they would eat slugs poisoned by pellets, but the decline was most evident on farmland. A victim of intensive farming practises and loss of hedgerows, they now seem to be making a welcome comeback. Parks and gardens are important habitat for this brilliant songster.

Getting Warmer

And what happened to winter? Milder winters must have helped the thrush that feeds mainly on invertebrates. In the past couple of years, in the centre of England we've had no more than a brief flurry of snow in January. I remember as a boy in the fifties and sixties struggling to school through the piles of snow thrown up on the pavements by the snowploughs, or building great forts of snow for epic snowball fights with the enemy from the next street, and sledging down Dawson's Field every year.

The last few years have seen record high temperatures both globally and in the UK. In Britain the effect seems to be that the seasons merge more, and we are expected to experience global wetting as much as global warming. Certainly we've had lots of serious floods of late. Maybe I'm tempting fate, and perhaps by the time you read this we will have had another 'Beast from the East' only more severe. But it seems to me that climate change is certainly happening, and accelerating, and is caused substantially by Homo Sapiens. Those who know what they are talking about predict that in the next hundred years global warming will seriously threaten the livelihoods of millions of people.

I remember back in 2007 reading the report of the United Nations IPCC, (the Intergovernmental Panel on Climate Change composed of the top scientists from all the member nations). Unsurprisingly it made front page news everywhere. It said the world is warming in direct proportion to the quantity of greenhouse gases in the atmosphere. The world was already 1 degree celsius warmer than in pre-industrial times, and the climatologists predicted at least a 2 degree rise in global temperature in the 21st century unless there is coordinated action to dramatically reduce emissions. The temperature rise would have serious impacts, such as more violent weather events, more drought areas and rising sea levels. They went on to point out that a 4 degree rise would trigger various 'tipping-points' with shrinking ice-sheets reflecting less sunlight, and melting permafrost releasing huge amounts of methane. A 6 degree rise doesn't bear thinking about, with the great forests burning out of control and mass movements of people fleeing inundated coastal cities or seeking water and food.

Could this armageddon scenario happen? Well, so far, all the predictions in 2007 have proved to be correct. The decade to the end of 2019 was the warmest on record with the highest ever recorded temperatures in the UK and Western Europe as well as elsewhere. There have been more serious storms and floods, forests in Australia have burned for weeks, and the ice sheets are melting at an alarming rate. Sea levels are slowly rising. No wonder Greta Thunberg and the world's children are protesting. I despair about the world's leaders not doing enough about this critical challenge, possibly because the four or five year electoral cycles mean unpopular and expensive decisions for the long-term can be postponed again and again.

We don't need the climatologist to show us the statistics. Even in a park in the centre of a town in the English Midlands, the evidence of a changing climate is there for all to see. The blossom on the wild cherry has been out since December. In January, a few daffodils are showing under the shelter of the copse. The birds are pairing up and nesting earlier. Is that so bad? Clearly other parts of the world will be impacted before the UK. Even so, many of our bird species are suffering already, particularly the seabirds which rely on fish that prefer colder waters. And some invasive species from the south can devastate our local wildlife. But it is the social-economic impacts of climate change that will have the most profound consequences. We will all be affected if global heating continues unchecked.

God's Messenger Passes By

Could this be a fateful omen? No, I think not. But it is still rather weird to hear the unmistakable croak-croak of a raven in the middle of a provincial town. That sound reminds me of holidays in the wilder parts of north-west Britain, the Welsh Mountains, Western Ireland or the Scottish Glens where the raven's croak would carry for miles, seeming to echo off the hillsides. I look up, and there it is, with diamond-shaped tail and heavy bill, flying purposefully over the park and ignoring the mobbing of two of the local crows. With its four-foot wingspan, it's a wonderful bird to see here.

The raven is deeply embedded in European folklore, and usually it is not good news. *Corvus corax*, to give its Latin name, is associated with death, mystery and foresight. In the Viking world, the raven was an iconic and magical creature; a powerful symbol of war. It was believed to be the eyes and ears of the chief Viking god, Odin. He was accompanied by two ravens, Huginn ('thought') and Muninn ('memory'). They flew all over the world, to bring information to Odin. In ancient Greece, as well as being a messenger of the gods, the ravens were associated with Apollo, the god of prophesy. And everywhere they are a symbol of bad luck, perhaps because they were always present at battlefields, feeding on the corpses. The collective noun for ravens is an 'unkindness' or in some areas a 'conspiracy'.

Well, I regard it as fortunate to see ravens in Warwick. Until relatively recently, they were extremely rare in central England. Now, like the buzzards that have spread from the more remote zones, they are occupying territory across the country, feeding on road-kill and benefitting from the resurgence of the rabbit population following the myxomatosis years. Ravens remain aloof, in urban areas only seen flying over, apparently just passing through on important business. But it's great to glimpse them, and to hear the echoing croak-croak of this mysterious bird in the centre of town.

A Rarity in the Reeds

And what's this? Another welcome surprise. I am walking round the reed-fringed fishing pools when I hear a sharp "chip". I wheel round. That's not a wren. I think I know what it is, but must find out for certain. I creep towards the sedge by the pool. There's something in there. I can see the reeds twitching. Then it appears on a reed stem, poses for a moment flicking its broad tail, then flies off. A Cetti's warbler, (pronounced chetty's).

In many ways this is an unremarkable bird. It is the classic SBB. For birdwatchers this means a Small Brown Bird which is very difficult to identify. But the Cetti's is not unremarkable. It is one of the few warblers that does not migrate away from the UK in winter. This makes it vulnerable to harsh winter weather, but the mild winters of recent years have seen it extend its range northwards. The Cetti's warbler was first registered as breeding in the UK in 1973, and it has spread along the south coast from Kent to Devon. It's a very elusive, fairly nondescript brown bird, but the male has an easily-recognised, short, explosive song-call in the spring, usually ending with, "What's my name? What's my name? Cetti-Cetti-Cetti".

Perhaps this individual will adopt St. Nicholas Park as its territory. Cetti's warblers have appeared in Warwickshire in recent years, notably at the Warwickshire Wildlife Trust's Brandon Marsh, but it's very good to find one in the centre of Warwick. Will it find a mate? We will see. My notes tell me this is bird species number 74 I have identified in the park since I started observing the wildlife here a couple of years ago. How many will there be by the end of the year? Will it be possible to reach 80 different birds? It's an ambitious target for the centre of a town.

But I'll hope for more surprises in this urban patch of riverside habitats. Otters have been seen in the River Leam, a tributary of the Avon upstream in Leamington Spa. I'm envious. How wonderful it would be to see these elusive wild creatures in the middle of Warwick. It's unlikely, but you never know.

February

*"February is a suitable month for dying. Everything around is
dead, the trees black and frozen so that
the appearance of green shoots two months hence seems
preposterous".
(Anna Quindlen, American writer).*

Great Spotted Woodpecker

The Sounds of Spring

Well, Anna Quindlen, the middle of England is anything but dead in February these days. On the contrary, everything is firing up with life's indefatigable spark of annual renewal. This February begins with some blustery weather, but it remains mild and by the middle of the month, an unusual southerly airstream from Spain brings a spell of extremely warm weather – 14 degrees and even warmer in the sun. The Cetti's warbler is now in full song in the reed bed, shouting its name. The dunnocks are jingling their pretty, trickling song from the tops of the bushes, and the robins are singing fit to burst. I can even hear a skylark in the far distance.

Bird-listening is as important as bird-watching if you want to know which species are around. For many people the various trillings and twitterings in the trees must be, at best, confusing, and at worst ignored. After a lifetime of watching and listening to birds I have learned to identify most calls and songs, though some of them are a bit tricky. It's a genuine pleasure to walk through the park, and know exactly what is around, and even what the birds are doing. It is the environmental soundscape that connects us with the natural world.

The Bird of the Chortling Stream

There is a stream that runs through the park. It is called St. John's Brook, because after running in a culvert under the main road to Coventry, it reappears in the grounds of St. John's House, an elegant Jacobean mansion now used as a museum, and the stream flows through the park between flower beds to join the Avon where there used to be a chain ferry across the river. A little wooden bridge across the brook is often used by the children for games of Poohsticks.

A bright 'chittick-chittick' call tells me there is a wagtail searching for insects along the brook. This is not the pied wagtail, seen in so many towns running about on the tarmac of the car parks. This is the grey wagtail, and it is a treat to see it in the town. It is a bird of upland streams. But in the winter some will move down to lower levels, but always where there is running water. Its cheerful call inevitably seems to have a backing track of a trickling and chuckling stream. Often mistaken for the elusive yellow wagtail, the grey wagtail has a beautiful lemon-coloured underside and a slate-blue back. Its long tail pumps endlessly. Why? I really don't know. But with its bounding flight and wagging tail, it always looks as though the grey wagtail is enjoying life enormously. It is a fabulous bird to find in our urban park.

The Stormcock

The blustery weather in early February has encouraged a mistle thrush to fly up to the top of a larch, and sing into the wind. Known in the countryside as 'The Stormcock', this is one of the very few birds that likes to sing in a strong wind. And its song is completely different from the sharp, confident, rinsing-the-ear brilliance of its cousin the song thrush. The song of the mistle thrush is sad. It produces wistful or mournful phrases, rather half-heartedly, in a way that suggests all is lost. This song does not represent the character of the mistle thrush at all.

Named after the mistletoe berries that it likes but that are poisonous to most other birds, it is a bold and aggressive bird when defending its territory. In the winter when the redwings and fieldfares arrive from alien lands, the mistle thrush chatters its threatening call and spends hours trying to defend its berry-rich hedges. At nesting time, the smaller song thrush tries to hide its nest in hedges and dense bushes, hoping that the predators will not find it. Not so the mistle thrushes. They say "Come on if you think you're hard enough!" and nest openly in the fork of a tall tree, defending it from magpies and crows with aggressive attacks and a loud churring alarm call that sounds like a piece of machinery. As a boy, I was innocently climbing a tree in the local woods when I was suddenly struck on the head by a chattering demon and nearly fell 30 feet. It had drawn blood. It was a mistle thrush. I didn't know it but I had climbed too close to its nest. Ever since, I've admired this bird's bold defence of its territory; but why is its song so miserable?

The Drummer in the Woods

Mid-February; the sky is blue and the wind has gone. The pools are like mirrors. Each weeping willow has its reflected counterpart seemingly reaching up to touch the drooping strands. It is St. Valentine's Day, when Geoffrey Chaucer believed that the birds would gather together to find a mate. In his poem, 'The Parlement of Foules', written in the late 14th century, he describes a dream in which he sees all the birds assembling on St. Valentine's Day before the 'noble empress' Nature. Translated from the Middle English, he relates his vision like this:

> *What can I say? Fowl of every kind*
> *That in this world have feathers and stature,*
> *Men might in that place assembled find*
> *Before the noble goddess Nature,*
> *And each of them took care, every creature,*
> *With a good will, its own choice to make,*
> *And, in accord, its bride or mate to take.*

Certainly most of the birds in the park are busy pairing-up in February. A sharp 'Quick...quick' from an oak tree by the pools tells me that a great spotted woodpecker is calling for a mate and she should be quick about it. Then suddenly he starts drumming; a sure sign of a woodpecker looking for a partner. Perched on a dead branch selected for its resonance, the bird sends out short bursts of sound that seem to echo through the trees. The unmistakable drumming carries long distances and is one of the typical sounds of early spring in woodlands in Britain. Both the males and females drum. I can see that this individual is a male, identified by a red patch on the back of his head.

You would think he would get a headache. It seems the woodpecker has evolved so that it can't suffer brain damage from this vigorous hammering. The great spotted that drums more frequently than other woodpeckers has been the subject of several research projects. In its drum-roll, it can deliver as many as 40 strikes per second, each one of them generating a force of 1,000g (where g is the force of gravity). Scientists found that to cope with these major impacts, a woodpecker has evolved a suite of modifications. As you might expect, these include a reinforced, extra-thick skull. But the skull also contains special spongy sections that act as shock absorbers, plus a cushioning layer of muscle underneath it. And a unique extended bone in the skull, the hyoid, helps to hold the brain in place like a harness. Sure enough, my individual in the top of the oak tree flies off with his characteristic switchback flight and chip-chip call, clearly none the worse for his head-banging.

And another species of woodpecker is becoming vocal in this warm early spring. Standing by the pools in the park, I can hear across the rough ground to the east of the line of trees the 'yaffling' of the larger green woodpecker. He is also calling for a mate.

And like its smaller relative, this woodpecker has evolved a unique skull structure, not just to protect its brain when it drums, which it does occasionally, but also to help it get to its favourite food; ants. To get at the ants, and even in the winter it can attack ants' nests, it has a very long, sticky tongue. This rests in a special curved slot in the top of its skull, and when it flicks it out it's nearly four inches long, a third of its entire body length.

The green woodpecker is a striking bird with its red and black head and green back, but is easily frightened off. I always think it looks slightly insane, with a pale staring eye, and on the ground an anxious posture with its beak pointing upwards as it constantly looks around suspiciously. And there it goes, making off with its looping flight and showing its lemon-yellow rump. The rather hysterical laughing call of the nervous green woodpecker is another sign that spring is on its way.

Not Sparrows

In the hawthorn and blackthorn bushes beside the pools there's some frantic dashing and chasing around by small groups of an undistinguished small brown bird – yes, another SBB. It could easily be mistaken for a female house sparrow, except that, unlike its noisy neighbours that come across the park in squabbling groups from the nearby housing estate, it is comparatively quiet, emitting a regular thin peep, rather like a soft blast on an old-fashioned policeman's whistle.

This is the dunnock, also known widely as the hedge sparrow. But it is not a sparrow. The dunnock with its finely patterned back and smudgy grey head has a more delicate beak than the seed-cracking sparrow. And it doesn't nest in colonies, but secretly in thick hedges and shrubs, hence its country name. It is a member of a very small family of birds called accentors.

Dunnocks are common in gardens and parks, usually creeping about mouse-like close to the ground, picking up the tiniest morsels. But in the breeding season they become extraordinarily randy, with both sexes continually flicking their wings to show that they fancy a bit of how's-your-father. But it is more a question of who's your father? The females seem happy to be inseminated by any male who comes along. A dunnock's nest, typically cradling 5 or 6 eggs, may have a different father for each one of them. For a bird that will spend its life in a small piece of territory, this means that the next generation will have a varied gene-mix when they are old enough to feel the urge.

Geoffrey Chaucer knew about the dunnock's promiscuity. His poetry shows how closely connected to their natural surroundings people were in the 14th century, when the population of England was only about 7 million with most living in the countryside. In the Prologue to his Canterbury Tales describing the pilgrims, one of his most despised characters is The Summoner, an official of the Catholic church – a sort of policeman - who summoned people before the ecclesiastical courts.

> *A Somonour was ther with us in that place,*
> *That hadde a fyr-reed cherubynne face,*
> *For saucefleem he was, with eyen narwe,*
> *As hoot he was and lecherous as a sparwe.*

In translation from the Middle English:

> *A Summoner was with us in that place,*
> *Who had a fiery-red cherubim's face,*

Carbuncled so, and his eyes were narrow.
He was hot and lecherous as a sparrow...

My guess is that Chaucer was referring to the Hedge Sparrow – the lecherous Dunnock.

The Fastest Creature on Earth

I am excited. As I am walking out of the park, a peregrine falcon cruises over the car park heading for the tower of St. Mary's Church. This is a rare sight in Warwick and a welcome one. The peregrine has a beautiful aerodynamic shape, with a muscular chest and sharply pointed wings that beat in a fast winnowing action. I can clearly see the black face-mask above the white throat. This king of falcons, much prized by falconers in the old days, is totally at ease in the blustering breeze. It is the master of the skies, able to stoop on prey at 200 miles per hour, making it the fastest creature on Earth.

Peregrines in the UK declined to very low numbers in the last century. They were shot during both World Wars because they could intercept pigeons carrying important messages from the Western Front. But much more serious was the widespread use of DDT on farmland after WW2, when there was an intensive drive to produce more food. This insecticide accumulated in the peregrines and other birds of prey that fed on farmland birds such as pigeons, and their eggs became infertile. Populations plummeted.

In recent years, peregrines have spread across the country, occupying sites in many cathedral towers. To be strictly accurate, St. Mary's in Warwick isn't a cathedral, it is a Collegiate Church that was run by a College of Canons rather than directly by a Bishop. It is a magnificent structure housing the celebrated Beauchamp Chapel, the resting place of the

Earls of Warwick. And it has a very tall tower. Peregrines have nested successfully for several years in the Town Hall at nearby Leamington Spa. Could they be showing interest in St. Mary's Church? It seems to have plenty of ledges, slots and indoor spaces to attract them. But so far peregrines have rarely been seen around St. Mary's. Perhaps they are put-off by the bells, the chimes of the clock, and the carillon that plays tunes four times a day.

Not Seagulls

Throughout the month there are large numbers of gulls in the park, white blobs across the playing fields all facing the prevailing breeze. They are black-headed gulls, though they don't have black heads during the winter, and to be accurate, when they do develop their breeding plumage, it's easy to see that their caps are chocolate brown, not black. But I suppose chocolate-brown-headed gull is a bit of a mouthful. In the winter their 'black' heads become a smudgy spot behind the eye.

I swiftly do a count. There are at least 250 on the playing fields. Many of them are 'pattering'. This is great fun to watch. They do a little dance on the spot. It seems that the worms that are a gull's staple diet during the winter are stupid enough to think it is raining and come to the surface to enjoy the moisture. The fools! It clearly works, as the dancing gulls all seem to be benefitting from a juicy meal, pulling out worms every few minutes.

My grandmother used to say when she saw gulls in the fields around her home in Yorkshire, "There must be bad weather at sea". The idea that gulls are 'seagulls' isn't quite right. Technically there is no such thing as a seagull. They are gulls, and though they are very common around our coasts, some spend their entire lives inland. The lesser black-backed gull nests on cliffs along the Atlantic coast and migrates quite long distances in winter, across the UK and as far south as West Africa. It is on the Amber List of endangered species because the UK is home to 40% of the European population and they are declining in many of their traditional nesting sites. But now they are becoming more urban, nesting on tall buildings, though they do seem to prefer being near the coast or large rivers where the pickings are easier. In the winter months, lesser black-backs appear in the park in small numbers quite frequently. The smaller black-headed gulls are extremely common in the park. They seem to be content to spend their lives inland. They often nest on islands in lakes and are frequently seen following the plough in the autumn.

The eggs of the black-headed gull have been prized by gourmets since Victorian times, and I was surprised to learn recently that they are still harvested in some parts of the country. Collecting gulls' eggs is tightly controlled. There are only six sites in England where harvesting is permitted, with about two dozen licensed collectors who are allowed to take only one egg from each nest. The licences are hereditary, passed down through families. The attractive sage green and speckled eggs end up on the breakfast tables of the affluent, and pound for pound they are pretty pricey – about £6 per egg! They are said to taste delicious. But the word 'gullible" comes to mind.

Walking by the river, I notice another, slightly larger gull among the black-headed crowd who descend screaming on the children feeding the ducks. The common gull is not well named. In the centre of the country it is far from common. But here it is, in our park. An intermediate size between the black-headed and the larger herring gulls and lesser black-backs, the common gull can be identified by its smudgy grey head in winter, and in flight by the black and white panels on its wing tips. Trying to identify gulls is always rather difficult, but the ubiquitous black-headed gull is relatively small with an agile flight, and its sharp wings have a white leading edge that is normally easy to see.

A Small Surprise

The beautiful clear weather continues with surprising warmth from a sun still fairly low in the sky. It is 23rd February, a Saturday, and the playground is full of children in light clothes enjoying the swings and climbing-frames. Beside the river, I'm surprised to see a butterfly fluttering past. It's a small white, often called the 'small cabbage white'.

It is not unknown to see a butterfly on a warm day in February, but they are usually the species that hibernate and have been awoken by the sun, even temporarily. So a peacock, red admiral or a brimstone might venture out of its winter hiding place on warm winter days. But the small whites, or 'cabbage whites', do not hibernate. They mate, lay eggs, then survive for only two or three weeks. The following year, the eggs go through five larval stages with the green caterpillars a menace to gardeners, then become a pupa before the butterfly emerges. This all takes about a month, so this individual fluttering by the Avon emerged from a tiny egg in mid January, showing that there have been unusually mild temperatures during the heart of winter.

Pollinators

In the following days more insects appear. A very large buff-tailed bumble bee drones past, then clumsily drops into some undergrowth, its wings clattering the blades of grass. It is almost certainly a queen looking for a hole to make her nest.

These common bumble bees often use a hole or tunnel made by a small mammal. Beatrix Potter knew this. Mrs. Tittlemouse was very irritated to find a bumble bee, (Babbitty Bumble), in her nice clean house underground.

There are an astonishing 270 species of bee in Britain, 25 of them are different types of bumble bee. The others are different species of solitary bee, plus the single species of honey bee. I confess I find bees a little difficult to identify. With bumble bees I tend to look first at the colour of their bottoms. The white-tailed and buff-tailed are pretty common. Also the red-tailed bumble bee is easy to spot. The common carder bee is a smallish one with ginger fur on its back, often seen in gardens. All the others I find rather challenging.

The population of wild bees in the UK has been in serious decline since 1980. Recently it appears to have stabilised, but according to a study by the Centre of Ecology and Hydrology in Oxfordshire, about a third of the wild bee and hoverfly species in Britain are in decline. Worldwide, wild bees are in serious trouble. Numbers have tumbled by more than 30 per cent in a generation. This is serious for us all. Bees are among the best pollinators in the world, underpinning food production. A long-term study by the University of Ottawa, published in 2020, concluded that bumble bee populations are declining in many parts of the world. And it found that climate change is a major driver of those declines as the bees struggle to survive in hotter and drier conditions in some countries. Habitat loss, diseases and pesticides such as the 'neonicotinoid' nerve agent, banned by the EU but widely used in the USA, have also played key roles. Not surprisingly, bee loss is most pronounced in the USA. Why they would risk this vital part of earth's ecosystem by licensing dodgy pesticides is a mystery to me.

In the park, the honey bees have started to emerge in the warm weather, prospecting in the copse at ankle height and climbing upwards into the oval pods of the snowdrop flowers. I wonder if 'snowdrop honey' could be a marketing success. Honey bee numbers are definitely declining in Britain. The main reason appears to be a nasty virus called 'chronic bee paralysis' that kills about 40% of the bee colonies it infects. The virus was confined to just a few areas of the UK in 2007, but a study by Newcastle University published in 2020 showed that it is now present across most of the British Isles. As if that wasn't bad enough for the honey bees, they are also affected by pesticides and habitat loss caused by urban development and intensive farming. It's reported that in the past 100 years, ninety-five percent of Britain's wildflower meadows have disappeared. So gardens planted with flowers, and parks with flower-beds and areas of uncultivated rough ground, have become vital places to sustain our important pollinators.

There are also clouds of dancing midges under the willows beside the pools, illuminated by slanting shafts of sun. So far it has certainly not been a bitter-cold winter in this region. By the end of the month, figures from the Meteorological Office show that the temperature in Britain was an all-time record for February. It was up to 21 degrees in London and close to that in the Midlands. Globally it's announced that the previous year was the second hottest on record. In Europe it was the warmest ever recorded. Earth's temperature over the previous five years was about 1.2 degrees warmer than pre-industrial times.

March

'March month of many weathers wildly comes...'.
(John Clare. The Shepherd's Calendar).

Kingfishers

Not So Wild

March is supposed to be a bit mad, with John Masefield's 'Mad March days' and the mad March hare boxing in the open fields or baffling Alice in Wonderland. These days the month of March can't claim exclusive rights to wild weather. The storms and gales from the West can come at any time of year, bringing floods more frequently. If anything, the month of March seems to be a little less wild than in former years.

In St. Nicholas Park in early March, it's windy enough with the tail-end of an Atlantic storm that has battered the Welsh coast, but it is relatively warm with no suggestion of frost or snow. I'm waiting for the first migrant birds to appear on these south-westerly airstreams. It is enlivening to watch and hear the start of the breeding season, with the air full of urgent birdsong.

The black-headed gulls dotted across the open grass suddenly all take off as one, swirling around in the breeze. Something has disturbed them. The peregrine? No, high above there is a buzzard cruising effortlessly into the wind. It is unlikely to go for a gull, certainly not in broad daylight, but the gulls are taking no chances. Buzzards are opportunists, taking carrion and sometimes lurking in trees ready to drop on to a rat or rabbit. But they can also stoop from height in a long angled dive with wings half closed, trying to surprise a squirrel, a pigeon or even a gull.

The reeds beside the pools are waving and rattling, but it is still easy to hear a bright little repetitive call; 'chink, chink, chizzuk'. And there it is, perched on the top of a hawthorn bush - a reed bunting. This male is in his smartest breeding plumage, with a black and white head, a streaked brown back, and a longish forked tail with white outer feathers that flash as it flies away with a characteristic bouncing flight.

Reed buntings tend to spend the winter on farmland with mixed flocks of buntings, yellowhammers, finches and sparrows. Now it's time to find a good nesting site in waterside vegetation. It's a very attractive bird to turn up at springtime in the local park. Perhaps this singing male will attract a mate, and she will build a beautifully formed cup of grasses and moss hidden at the base of the reeds.

Predators

The river is rather choppy in the breeze, and there in the middle, bobbing up and down, is a very early brood of tiny brown and yellow ducklings, thirteen of them, probably less than a day old, with their mother leading them across to where the children feed the birds. The drake mallard is an absentee father. Bringing up the kids is left entirely to the female. It is sad to know that most of them won't survive very long. Crows, magpies, gulls and even pike will take ducklings when they are small.

And immediately I can see this family is in danger. A large lesser black-backed gull is circling overhead. I feel like shouting, 'Look out. Stay closer together'. Too late. The gull banks down suddenly, picks a duckling off the surface with its powerful yellow bill, and veers away on the breeze. The remaining ducklings follow their mother to the shelter of the bank. She seems unperturbed. Did she even notice this brief murderous raid? I am glad to say the little girl throwing crumbs of bread clearly did not see it. I think she might have been seriously upset. But this is nature's way. Wildfowl have large families. Only a few of the young ones will survive. The predators must also eat and feed their own young.

There are plenty of predators threatening ground-nesting birds like the mallard. The fox is a major danger at night, but I have also occasionally seen a brown rat nosing about at the edge of the river, and one swimming quite assuredly all the way across. And there are other dangers for the wildfowl and moorhens as the breeding season arrives.

The Weasel Family *(Mustelidae)*

A friend from the Warwick Natural History Society has seen a mink crossing the path ahead of him near the pools. That is bad news. Released from fur farms by animal welfare campaigners in the 1950s and 60s, the American mink has spread across the country and is now a menace along Britain's waterways, devastating water vole populations. Its smaller native cousins in the *mustelidae* family, the weasel, stoat and polecat, are also a threat to ground-nesting birds. So the female mallard must rely on camouflage and remain perfectly still on the nest, probably holding her breath if a predator is close by!

And here's evidence of another member of this family of omnivores. There is a bank of grass at the far side of the Kingfisher Pools, neatly mowed by the contractors who maintain the park. This morning it does not look so neat. An area about six meters square has been grubbed up with the turf turned over in ragged lines. I have a close look and there are several slimy, black droppings, and some tell-tale paw-prints with the impressions of claws. These are the unmistakable signs of badgers grubbing up the grass to find earthworms and other juicy morsels. But badgers also enjoy a nice egg for breakfast. They are yet another threat to ground-nesting birds in the breeding season. Even so, it's good to know there is a sett somewhere nearby, and that under cover of darkness they venture into the park to dig for worms.

Pigeons

In my view the pigeon family is often underrated. The ubiquitous wood pigeons with their flush of pink on the breast and white wing flashes are already vocal in the trees. The cooing of the wood pigeon is a classic sound of a British summer. Mellow and repetitive, it seems to evoke lying in a hammock in the shade of a cedar, with a romantic novel and a glass of chardonnay to hand. There are many interpretations of its 5-note call. 'Bring two cows, Paddy' is a popular version in Ireland, and in Wales it's, 'Take two cows, Taffy". Here in early March they are cooing their soothing song to tell us that it will soon be summer.

In the park there is another similar call in the spring. This has three notes rather than five, and to be honest it is a bit of a repetitive drone. It's the collared dove, a very elegant soft grey and lilac dove with a black and white tail and a black stripe across its neck that gives it its name. Some say the three-note call makes it sound like a football fan: 'Un-it-ed, Un-it-ed, Un-it-ed!' but chanted without much confidence that United will come back from two goals down.

The collared dove is an invasive species. There was excitement among birdwatchers when the collared dove nested for the first time in Britain, (in Norfolk in 1956). Within 20 years it had colonised every part of the UK including the Outer Hebrides. Wherever there are human settlements the adaptable collared dove will thrive. Normally they nest in fir trees, but many are happily converting to ledges on buildings in towns. One pair chose a bowl of bulbs on a high ledge at Leeds University and sat tight as the hyacinths grew around them. Some have made their nests entirely of bits of wire when no twigs were handy.

In the park they are quite easy to distinguish from the wood pigeons, quite apart from the droning call. The male wood pigeons show off in March by flying upwards with their wings clapping, then gliding down with the wings held out stiffly. The collared dove also performs a display flight, but with the wings bowed downwards, rather like the curve on a paraglider's chute, and usually uttering a rasping 'kaa – kaa'.

There are two other pigeon species regularly seen in the park. First, let's dismiss the feral pigeon, the 'London pigeon' now banned from London landmarks. This bird was domesticated many centuries ago, to provide eggs and squabs, and to exploit its homing instincts for communication. It was bred from the rock doves that live on coastal cliffs. In Warwick there is a colony of feral pigeons living under the bridge over the Avon, but since the peregrines have started to appear in the area, they seem to have abandoned the church steeples and the castle battlements.

The other member of the pigeon clan found in St. Nicholas Park is the stock dove. This is something of a surprise. The stock dove is a bird of the open country, seen in small flocks on stubble fields. More compact than a wood pigeon, it is the smart little cousin with an iridescent bottle-green patch in its neck and a pink chest. Here in the park, a pair has taken an interest in the weeping willows directly above the spot where families feed the ducks, swans and gulls. Unlike the wood pigeons that build flimsy nests in all sort of places, some of them stupidly placed and quickly abandoned, the stock doves nest in holes in trees or walls. The old willows by the Avon have a few inviting holes, so let's hope they breed successfully.

Buzzards

In the second half of March, a ridge of high pressure has settled over the UK, and the weather becomes remarkably warm.

After the wind and rain, it is flat calm with blue skies and cotton-wool clouds. Soon the migrants will arrive.

The buzzards seem to love soaring in the sunshine. I hear their mewing cries, and look up, squinting into the brightness. There in the blue are two pairs having a territorial dispute, but fairly casually, as though the border protocol is already agreed. As they cruise in circles, every now and then one closes its wings and falls like a stone before spreading its wings and swooping up again as on a roller-coaster. The males are showing off, telling their rivals, "Don't mess with me", and telling their mates, "Look how fabulous I am".

The buzzards have spread into England from the remoter parts of the UK in just a few years. Their staple diet is rabbit. After the myxomatosis years, the rabbits are back everywhere, and so are the buzzards. I remember clearly New Year's Day in 1987. I was sitting in my study overlooking the Warwickshire fields, and as I glanced out of the window there was a buzzard wheeling over the garden. I nearly fell off my chair. A buzzard? Just 12 miles from Birmingham? Now they are a common sight across the UK. And it is a beautiful bird – a mini-eagle with a characteristic outline in flight – broad wings with fingered tips – and a plaintive cat-like call that seems to be carried by the wind.

The Migrants Start to Arrive

On the 18th March, the first migrant warbler finds its voice in the park. It's a chiffchaff, saying its name in the ash trees by the pools. Other European languages have similar names for the chiffchaff: The Dutch use "tjiftjaf", Germans use "zilpzalp" and the Welsh say "siff-saff". This bird in the park might be one of those that has stayed in the UK all year, because as the winter months become less harsh some decide to hang around.

But it is much more likely to have flown all the way from the Mediterranean or West Africa to breed in a park in the middle of England.

The chiffchaff with its soft grey or yellowish plumage is almost indistinguishable from the willow warbler, but if you get a good view, look at the legs. The chiffchaff has black legs and the willow warbler has flesh-coloured legs. But it's the song that separates them, as Gilbert White observed in his wonderful 'Natural History of Selborne'. He decided they had to be different species because their songs were so different. The willow warbler has a trickling wistful song that runs down the scale and peters out rather mournfully.

Three days later there are chiffchaffs calling all over the park and working their way along the willows overhanging the river. There has been what's called a 'fall' of migrants arriving in large numbers on a south-westerly breeze. And now there is a blackcap singing in the hawthorn bushes. Like a few of the chiffchaffs, this individual might have over-wintered here. Some do. And the blackcaps that turn up in our gardens in cold weather are often winter visitors themselves, moving to the more temperate climes of the British Isles from central and southern Germany. But the odds are that this singing male has flown in from Spain or North Africa. His liquid, halting song is a delightful confirmation that spring has arrived, and reminds me of holidays in the South of France, where the song of the blackcap in summer is ever present. The blackcap is a slim, grey warbler. The male wears the black beret; the female a ginger-brown one.

In the next few weeks of the spring migration, between 15 and 20 million birds will arrive in the British Isles to breed after some extraordinary journeys. The tiny and fragile warblers may have set off from Southern Africa in early February, managed to cross the arid Sahara, and braved the Atlantic storms to find safety in our temperate climate, possibly in a park in Warwick. At the same time, millions more birds are leaving our shores. The 'winter thrushes', the redwings and fieldfares that came here in the autumn to escape the sub-zero temperatures in Scandinavia, Russia and Iceland, are on their way back to their breeding grounds. Many starlings are heading for Eastern Europe to breed.

Migration Mysteries

A surprising number of our British birds migrate – about half of them according to the RSPB. Some species cover mind-blowing distances. The arctic tern is the record-holder, travelling from the south pole Antarctic ice shelf to breed in the far north, as the earth tilts on its axis towards the sun, giving them all-year-round summer. To gain the benefit, they fly a round trip of about 45,000 miles each year.

The migrants arriving in the centre of England haven't come quite so far, but a trip from South Africa or the Congo rain forest is quite a feat, especially for the tiniest birds. But how do they find their way? It has been a question that has intrigued ornithologists for very many years. Gilbert White, the curious and observant 18th century vicar of Selborne, reckoned swallows must hibernate during winter in the mud at the bottom of pools. The notion that they could fly to Southern Africa, and that the fledglings could do it without parental guidance, and that they would fly back again the following year to nest in exactly the same nest site was - of course - preposterous.

But it's true. Scientific studies in recent years conclude that there are several navigation systems at play. Many birds seem to follow landmarks such as river valleys or coastlines. But this isn't always possible; many birds migrate at night and in poor weather, and there are no landmarks when they are crossing deserts or oceans. It is now established that birds can detect the earth's magnetic field. This is called 'magnetoreception'. There are two mechanisms that allow this. Birds have 'magnetite' in their beaks, a mineral that responds to magnetic fields. Their eyes also contain a light-activated protein called Cry4 which helps them 'see' the earth's magnetic field. They may also be using their sense of smell, (in some species this is quite acute), but the researchers aren't sure about that, and all-in-all the wonder of the annual mass migration of birds is still something of a mystery.

Some British species move relatively small distances to find the right habitat for nesting. On March 22nd, I walk into the park and immediately see that the black-headed gulls have gone. The day before there were at least a hundred spread out across the grass. It's interesting that they seem to have made a collective decision to go, and may even be on their way as a group to an established breeding ground. Not far away in the parklands of the Packington Estate, there is a bird reserve with lakes and a large colony of black-headed gulls on some of the islands. In Warwick, only a couple of the larger lesser black-backed gulls remain on the abandoned grassland.

A Large Rarity

Walking back along the riverside path I can see a large bird with slow wingbeats flying towards me past the walls of the castle and quite high over the road bridge. A heron, I reckon. There's one seen regularly along this reach of the river. But though this is the same size and roughly the same shape as the grey heron, this bird is white. It is a great white egret, the first I've seen in the park, and indeed only the third or fourth I've ever seen in Britain.

I have excellent views through my binoculars as it flies past quite slowly, heading upstream against the breeze. The legs are particularly long and I notice that it has black feet, unlike the more common little egret that has bright yellow feet.

Three types of egret have been moving into southern England from continental Europe in recent years. The little egret first bred in the UK in Dorset in 1996, and is now a common sight on estuaries and wetlands along the southern and eastern coasts all year round. It is a splendid little heron with plumes on its head and back, and an energetic style as it wades through the shallows trying to surprise a fish with its darting dagger bill. I had been hoping and half expecting to find one at the fishing pools in the park before too long as the little egret expands its territory into the midlands.

So it was quite a surprise to first see its much rarer and larger cousin, the great white egret. This beautiful member of the heron family is widely distributed across the tropical and warmer temperate zones of the world, but is very uncommon in Britain. It was first recorded as breeding in the UK in 2012 when a pair nested on the Somerset Levels. At time of writing there are thought to be only about 20 breeding pairs in Britain, so it was a privilege to see one cruising through Warwick. It is bird number 74 on my park list.

The third egret colonising Britain is the cattle egret. This is a stubbier, white and yellow bird that gets its name from its habit of associating with cattle to pick up the insects disturbed by the livestock. In Africa they are often seen riding on the backs of water buffalo. The lumbering animals seem pleased to have their personal groomers picking flies from their eyelids and nostrils. In Britain this egret too has started to breed in small numbers along the south coast. The increase in numbers of these elegant invaders with their striking white plumage is another sign that in the past twenty or thirty years, our climate has been changing, with hard winters less frequent, and warmer and wetter weather throughout the year.

Fishing

The weather in late March is now glorious, with chocolate-box skies of clear blue and cotton-wool clouds. The magnolias beside the brook are covered in soft pink and white blooms standing vertically at the end of each twig, making the trees look like candelabras. The long strands of the weeping willows along the river are bright lemon yellow stippled with pale green spots, the colour of their leaf buds. The long stems of the pussy willows are covered with soft pale-yellow catkins. It's rather strange to find sticks of pussy willow in tubs in the local supermarket, but in this part of the country they are a traditional decoration in homes and churches around Palm Sunday.

In sheltered spots beside the pools, the blackthorn has been in flower for a couple of weeks, with its creamy-white blossom adorning the branches - another sign of an early spring. Conventionally the blackthorn is in flower in April and the hawthorn in May, hence its common name as the May Tree. Even the hawthorn is coming into bud and there is a flush of bright green along the hedgerow as its leaves start to unfurl in the sunshine.

Arriving at the river I hear a short, sharp 'peep' almost like a dog whistle. It must be a kingfisher, and sure enough, there it is, zooming along over the water close to the far bank, in a straight line almost as though it is on rails. The kingfisher is smaller than most people would expect, smaller even than a starling. It's good to see one here. Kingfishers are very vulnerable to harsh winters. It's estimated that during the very severe winter of 1963, when inland waterways were frozen solid for weeks, ninety percent of kingfishers in the UK died. It took many years before their numbers recovered.

And what's this? A second kingfisher zooms past and heads for the reeds in the aptly named 'Kingfisher Pools'. It would be great if a pair was to take up residence. I backtrack to the pools, and there it is, posing on a bullrush beside a sheltered pool near the footbridge. It is a male, with an all-black bill. The female has an orange lower mandible. With a splish he drops into the pool and emerges with a minnow, which he stuns on a branch, then juggles it so that it he can swallow it head-first, and gulp, it's gone. This is repeated successfully several times in about twenty minutes. It's clearly easy pickings. Then he is off, flashing his bright electric blue rump as he heads down the river.

You might wonder how the kingfisher can digest whole fish, bones and all. Like owls, herons and cormorants that also swallow their prey whole, kingfishers cough up pellets – pale grey capsules of fish bone and scales. Their nest, at the end of a tunnel dug into in a riverbank, accumulates regurgitated fish bones when they are feeding young, making them pretty foul nurseries. The male kingfisher will tempt the female with gifts of fish, flipping them round so that she receives them head first while fluttering her wings in what looks like a rather coy fashion.

In the following days, the kingfishers appear regularly along the river and by the pools. They are a delight to see, with their exotic plumage of orange breast and dazzling blue-green back. I am pleased that sometimes dog-walkers see me with my binoculars focussed on the reeds and they stop to ask, "What are you watching?" "Kingfisher", I reply nonchalantly, and they will say, "Wow, I've never seen one." So I show them through my binoculars and the world is suddenly a better place.

April

'Oh to be in England now that April's there'.
(Robert Browning)

Peregrine Falcon

The Awakening

For centuries writers have lauded the month of April in Britain as the great spring awakening, when flowers appear magically everywhere, the trees are coming into leaf, the birds are singing fit to burst, and the winter months are banished, at least for a few months. Chaucer's Prologue to The Canterbury Tales famously begins with April energising the pilgrims to go to Canterbury to renew their own lives. This a translation of the opening lines from the original Middle English:

> *'When that April with his showers sweet*
> *The drought of March has pierced root deep,*
> *And bathed each vein with liquor of such power*
> *That engendered from it is the flower,*
> *When Zephyrus too with his gentle strife,*
> *To every field and wood, has brought new life...*
> *... Then people long to go on pilgrimage'.*

This year there are no April 'showers sweet' in central England. The month begins dry, with light cloud and occasional patches of sunlight. It's quite chilly. Nonetheless, the mute swans are busy building a great mound of a nest beside the pools, song thrushes are in full voice, and there at the top of a wild cherry festooned with pink and white blossom, a blackbird is trying out his fluting fruity notes.

Blackbird

One of the nation's favourite songsters, the blackbird has an easy style with lilting, liquid tones, each phrase ending in a higher-pitched flourish. It is very unlike the urgent sharpness of the song thrush. If a blackbird sets up his territory in your garden you will be fortunate. He will sing through the summer until mid-July, sometimes before dawn, as Paul McCartney knew when he wrote of the blackbird singing in the dead of night. They also seem to like singing in the rain when other songsters fall silent, and they frequently sing immediately after a shower.

In an alder by the footbridge in the park, another male blackbird strikes up his song. But this bird looks different. It has dappled white feathers, white flashes on its wings and a pure white rump that shows brightly when it flies off. It's such a striking bird that walkers are often seen pointing it out curiously. It is a leucistic blackbird that has lived by the bridge for several seasons.

Leucism is a genetic condition causing a lack of the melanin pigment that gives plumage its colour. The condition is not the same as albinism, which is a mutation that prevents the affected bird producing any melanin pigment; so it is pure white with a pink eye. Albino birds are pretty rare, partly because their eyesight is thought to be impaired, but also because they are easy to spot and at a higher risk from predators. The same might apply to our leucistic blackbird, but he seems to be doing alright. He already has a mate keeping him company and searching for nest materials, and locals say he's been there for years. He could keep breeding successfully for some time.

The oldest recorded blackbird was one that had been ringed 20 years before. Whether this black and white blackbird's offspring will also have some white feathers is difficult to say. This genetic condition isn't necessarily passed on, though it can skip two generations.

On April 3rd there are two swallows feeding over the pools and the river – the first of the year. They are unlikely to linger long. They will be fuelling up before continuing the last leg of their migration from Southern Africa to the nest site they left the previous autumn. There are very few suitable nesting places for swallows in the town. They nest under cover on high beams or ledges, in a barn or stable where there is always an opening. Every year it is a particular pleasure to see the first swallows. Aristotle wrote, 'One swallow might not a summer make', but I reckon it certainly means spring has arrived.

The following day two house martins are catching flies over the river. These smart little dark blue and white birds, Shakespeare's 'temple-haunting martlet', don't need barns to shelter their nests. They construct their round weatherproof homes from tiny balls of mud mixed with saliva and glued beneath an overhanging roof. For more years than anyone can remember, there has been a nesting colony of house martins on the 17th century Market Hall Museum in Warwick market place, swooping over the shoppers at the food stalls with their cheerful 'chirrup' call. The early birds in the park will be moving on further north, with the Warwick colony starting to nest in May. They will be followed by the migrants that arrive from Africa last of all – the swifts.

Unusual Flowers

Suddenly spring flowers are all over the areas of woodland, scrub and grass that are left wild by the park rangers. There are still plenty of bright yellow daffodils, though they are past their best by April, and under the trees there are small mats of pale yellow primroses. Clumps of dog violets have appeared among the leaf litter in the copse, along with the golden stars of the lesser celandine. And here and there a wood anemone has pushed its way up and opened its soft, white, blush-pink flowers. I am not a botanist, but I am an anthophile – a lover of flowers – and have learned to identify quite a few of the flowers to be found along the river, beside the pools and under the trees in the park. I confess I have had to look up a few of them. But why not take a photo, and then use the internet to identify what you have found? A small blue and white flower among the violets turns out to be called Glory-of-the-Snow. No snow this winter, but it's still a glorious little flower.

By the Kingfisher Pools, the marsh-marigolds are easy to identify. Also known as kingcups, the large clumps growing right at the water's edge are crowned with brilliant golden flowers like large metallic buttercups gleaming in the sunlight. A few paces from the rushes by the pool is a quite tall plant with a cluster of white bell-shaped flowers hanging down and with lime-green spots on the end of each petal. It looks a bit like an outsized snowdrop, but this is the 'summer snowflake', *(leucojum aestivum)*. It's delightful. I take a close look – always a good idea with wild flowers. The six petals on each flower are finely veined, and the pollen on the anthers inside the bell is bright yellow.

Not far away there are several delicate pale lilac flowers on slender stems rising from a cluster of leaves. This is the cuckooflower, also known as lady's-smock, apparently because each petal is shaped like a milkmaid's smock. They sometimes appear along roadside verges at this time of year, but wherever they are found they need soggy ground to thrive. The name cuckooflower comes from the notion that the flowers appear at the same time country folk will first hear the cuckoo, as in the old nursery rhyme:

> *Cuckoo, cuckoo, what do you do?*
> *In April I open my bill,*
> *In May I sing night and day…*

Well, I should be so lucky! I haven't heard a cuckoo at all in Warwickshire for the past few years and certainly the unmistakable call hasn't been heard in the centre of the county town for some considerable time, according to members of the local Natural History Society. This is a species in steep decline. The British Trust for Ornithology reckons that since the early 1980s, cuckoo numbers in the UK have dropped by two-thirds. They readily admit, 'The reason for this decline is not known', but one theory is that their favourite 'host' species – the dunnock, meadow pipit, pied wagtail and reed warbler – are nesting earlier because of climate change, and the cuckoos arriving from Africa miss their chance to plant their eggs in these nests. But to be honest, no one has worked out the decline of the British cuckoo.

My own theory, based on not very much, is that the journey from Africa is becoming more hazardous because of increasing areas of arid land caused by climate change. Recent research using radio tagging of migrating cuckoos shows that there are two routes from sub-Saharan Africa to the UK. The Eastern route takes the birds across the Mediterranean from Libya to the toe of Italy, then through France to Britain. The Western Route comes through West Africa and crosses into Spain at the Strait of Gibraltar. This is the route with more fatalities as the birds head for their breeding grounds in the British Isles. Why the western route appears to be more dangerous is far from clear. The decline in cuckoo numbers needs more research. In the meantime, the cuckoo's 'wandering voice' as described by Wordsworth, and embedded in English folklore, is heard less frequently, and is much missed among the soundscape of April.

A Rare Raptor

April 9th will turn out to be a date I remember. It is a cloudy and breezy day. Having clocked-up 32 different bird species on my walk through the park, including 5 swallows and 2 house martins feeding over the river, I am walking through the churchyard of St. Nicholas Church when a large raptor drifts past the spire. With long wings and a longish tail it is clearly a kite. "That's really nice!", I say to myself. I have seen a red kite over the park only a few times before and this one is very close.

Red kites were common in Shakespeare's time, circling over London and other towns, picking up scraps. They were persecuted for years until only a tiny population in Wales remained. A reintroduction programme in various parts of England has been a spectacular success. Red kites are now a common sight soaring over many areas of England and Wales.

Between 1989 and 1993, ninety birds were released in the Chilterns area of outstanding natural beauty to the north-west of London. Ten years later 140 pairs were breeding there. Feeding largely on road-kill, they have multiplied since, and year by year are moving up the M40 corridor towards Birmingham. They are commonly seen in Oxfordshire, so Warwickshire is next for the arrival of the red kite. It's a really striking bird with pale grey and ginger markings and a long forked tail that twitches and turns in the wind.

But on this April morning standing by St. Nicholas Church, I quickly see this bird is not a red kite. The light was good and I could see that it was dark brown across the back with few markings, and the forked tail was also dark and not so long as the gingery tail of the red kite that is wiggled about in a characteristic way. This was unmistakably a black kite.

Gosh. I have often seen them when travelling abroad to warmer climes. In India, the Middle East, Africa and parts of mainland Europe, the black kite is common, cruising over Cairo or soaring over the Bosporus. But in the middle of Warwick? The birds that breed in mainland Europe have migrated from sub-Saharan Africa in the spring. Had this individual over-shot and gone too far north?

This bird drifted quite low across the car park. Despite its five-foot wingspan, it went unnoticed by the mums unloading their children from their 4 x 4s, and it headed south across the river with a couple of crows in rather half-hearted pursuit.

"Now don't get too excited", I tell myself. Warwick Castle has a daily display of their captive birds of prey called 'The Flight of the Eagles'. It is spectacular, with large vultures, including a beautiful lammergeier, and various small eagles skimming the heads of the crowds to take the baited lures from the keepers. Was this bird one of theirs, just having a cruise round from the castle? As soon as I get home I email the castle. "Do you have a black kite?" The same afternoon comes back the reply. "No, we definitely do not have a black kite". OK!!

There is a website logging black kite sightings in Britain and 'Birdline' records rarities seen in the UK. There are not many sightings of the black kite in the UK, but a dozen or so birds tend to appear in late March and early April, usually along the south coast. I enter my sighting. It seems an individual was seen the following day in Sussex, heading south. It may have been the Warwick bird that had realised it was heading in the wrong direction and had turned back towards mainland Europe. On my list of bird species seen in the park, the black kite comes in at number 75. I wonder how many more I will see or hear before the year is out.

A few months later, an article in 'Bird Watching' magazine quotes David Tomlinson of the British Trust for Ornithology predicting that the black kite will breed in Britain for the first time within the next ten years as our summers get warmer. As far as I know the bird I saw in St. Nicholas Park was the first recorded sighting of a black kite in Warwickshire. It will not be the last, if the summer temperatures continue to rise.

A Pair of Peregrines

In the middle of the month it is suddenly much warmer, with a southerly airstream bringing balmy weather up from Spain. More butterflies are emerging. There is a male orange tip dancing along the long grass by the river's edge. Like the other members of the white family of butterflies, the orange tip does not hibernate, but every spring it is a new generation hatching from eggs and pupating before flying. The male lives up to his name with large bright orange wingtips and is easy to identify. Perhaps the female should be called the black tip. She can be mistaken for the common or garden cabbage white or the green-veined white as she search for garlic mustard or lady's smock where she will lay her eggs, one at a time, on the underside of the leaves.

Now here are a couple of distasteful fact about this pretty butterfly. The orange patches on the wing tips warn predators that the orange tip tastes rather horrible. And the caterpillars are cannibalistic – they will eat each other if they come into contact. Now stop that, children!

A peregrine falcon is soaring in the clear blue sky over St. Nicholas Church, then glides away across the castle grounds. Perhaps one of the pair that has nested in Leamington Town Hall for the past few seasons. There is a webcam on the nest and they are incubating four eggs at the moment. It's brilliant to be able to see them in close up in real time.

The next day a peregrine is perched on a ledge on the St. Mary's Church tower calling repeatedly. Then suddenly there are three falcons zooming round the tower. It looks like two females, one carrying prey, and a smaller male who is making most of the noise, squeaking and chattering like a rusty pulley. All three make off at speed across the centre of town. So what was that about? It had all the signs of a territorial dispute, with a pair that had settled on the church tower seeing off an incursion by a second female. This is an encouraging sign that a pair has adopted the tower of St. Mary's and we may have breeding peregrines in Warwick for the first time.

Later in the month, both the male and female peregrines can be seen occasionally perched on the tower, favouring the ledge that supports the medieval shield of the Earls of Warwick. The earls may well have gone hunting with peregrines on their gauntlets. The 'Book of St. Albans' on Heraldry and Falconry, written in 1486, categorises the birds of prey in a 'hierarchy of hawks' that can be used for hunting by the various social classes. At number 15 on the list is the kestrel – for a Knave or Servant. At number 1 – for an Emperor – is the peregrine.

Lovely Long-tailed Tits

By way of contrast, on the same day that I have been watching the masters of the air competing for territory in the broad expanse of the April sky, I have also been watching some of our smallest and cutest birds at close range. Beside the pools in the park, a pair of long-tailed tits is collecting feathers. These are ridiculously pretty birds, with a striped head and pink-flushed plumage, and tiny beaks for picking up the smallest insects, aphids and spiders. The eponymous long tail makes the long-tailed tit look like a ball of wool with a knitting needle in it. When they fly from tree to tree, they bounce through the air, nearly always emitting their community call of 'see-see-see'.

Today this pair is emitting its soft 'zut-zut' call, which seems to have something to do with pairing up. One of them flits down to within a couple of paces of my feet to catch a small white feather. It disappears with its prize into a thick tangle of bramble. There it will be constructing a beautiful nest. The long-tailed tit's nest is like a small rugby football, made of lichen, moss, fine grasses, cobwebs, and as many as two thousand tiny feathers. There is a small entrance hole near the top. Inside it must be cosy to say the least, becoming rather overcrowded as the youngsters grow. A dozen crammed inside is not unusual.

But unfortunately these nests are often found by predators. Research shows that four out of five long-tailed tit nests will be predated. It leads to an unusual behaviour, where adults will help to feed a neighbouring brood if their own nest has not survived. Certainly the long-tailed tit population in Britain is thriving. A recent survey by the RSPB showed that their numbers have risen by 77% since 1973.

As well as being more communal than most, this very attractive little bird is easy to watch. It seems unconcerned if you approach and stand at close range. Some might say it is 'tame'. No, it is a wild bird, but it is – to use the naturalists' word – 'confiding'. The long-tailed tit is not frightened away if you stand close. Maybe it regards humans like cows – large, slow and harmless. Robins in the UK are famously confiding, but in continental Europe they are not, staying away from humans, and certainly not perching on Francois' garden spade or swooping down at his feet as he plants out his onions. Why this is I could not say, but it's a fact that more birds are confiding, meaning relaxed in human company, in the British Isles than in mainland Europe.

Confiding

A couple of days later I find myself watching another confiding bird, but this one is more difficult to see because it is more discreet and better camouflaged than the long-tailed tit. Walking along the riverside path in the park I hear a very soft 'sip sip' coming from a wild cherry. I think I know what that will be. Sure enough a treecreeper flies out to latch on to the bottom of the next tree along the path, flashing its bright white underside, closely followed by a second bird.

I think the treecreeper is particularly attractive. Its pure white chest is hidden when it hugs the tree trunk, and its mottled back blends in with the bark perfectly. Its long sharp claws and stiff supporting tail mean it can walk up trees, even on the underside of branches. Unlike the nuthatch that seems to enjoy being upside down, the treecreeper always flies to the bottom of a tree and works its way upwards, mouselike, before flitting to the base of the next tree. Its beak is a very fine curved instrument that can lever open the bark to reveal insects, spiders and woodlice even in the harshest of winters.

The treecreeper is certainly confiding. You can stand within six feet and it won't be disturbed. In our previous home, treecreepers nested in a fissure in an old oak tree beside our patio. I could stand right next to the tree, watching the parents going in and out to feed their brood. This is a lovely bird but not an easy one to spot, because it is small, is brilliantly camouflaged when working its way up the trunk of an oak or an alder, and its quiet call is at such a high register than some of us will not be able to hear it.

More Warblers and Butterflies

By the end of April, the wildlife in the park is literally buzzing, with more bees in evidence. And there are a few bee-flies zipping from flower to flower and hovering while they feed. These little balls of brown fluff are flies that look like bees. They have only one pair of wings; that defines them as members of the enormous order of flies, *diptera*. It's interesting to watch the bee-fly at work. It has an extremely long proboscis that penetrates deep into the flower, making it a valuable pollinator.

It has been record-breaking Easter weather, with temperatures in the midlands of nearly 25 degrees. Towards the end of the month it has become cloudier, but it is still very mild. After a few days at home with a heavy cold, I venture out into the park to blow away the cobwebs, and clock up 28 different bird species, including stock doves cooing, the Cetti's warbler shouting its name, blackcaps warbling, and several chiffchaffs saying chiffchaff to each other. So far there's no sign of its more reclusive relative the willow warbler, but an even more reclusive and much less common warbler makes a surprise appearance. To be accurate it doesn't appear. It remains stubbornly hidden among the scrub in the rough field beyond the pools while emitting a continuous faint metallic trill. It is this bird's extremely unusual voice that tells me that a grasshopper warbler has arrived from Africa. Its continuous insect-like trill sounds to me like a buzzing electronic fault. Others compare it to the sound of a fishing line being reeled in, which is why this unique song is called 'reeling'. I haven't heard a grasshopper warbler for several years, so it is a thrill to know that this is another unusual bird inhabiting the park, and it becomes number 76 on my park list of bird species.

In the past few days butterflies have emerged in some numbers, and are flitting along the grassy river banks and over the reed beds. Many are orange tips and in the brambles are several small tortoiseshells. When they settle, it's possible to creep quite close to these butterflies to have a good look. They have striking black and orange tiger-stripes on the forewing, and when they spread their wings, you can see that the trailing edges of the lower wings are attractively fringed with pale blue spots. The smaller common blue is less approachable, flitting away if you get too close. Only the males have the beautiful sky-blue wings, and when they settle, they seem to disappear as the camouflaged under-wings conceal the blue.

And flying past quite quickly in the sunshine are some larger, darker butterflies – looking almost black against the shining river. The peacock is named after the spectacular eye spots on each of its four wings, similar to the markings on a peacock's tail. When at rest, if it flicks open its wings, it may startle a predator just long enough to escape.

Much the same size, with a two-inch wingspan, is the fabulous red admiral. It hasn't got any connection with the navy as far as I know. Originally it was called the red admirable, but that was clearly too much of a mouthful. It is unmistakable, with white spots and vermillion stripes on the back of its black wings, and it's a remarkably friendly insect. If you stand next to a nettle-bed, the red admirals' favourite territory, one may even settle on your sleeve or hat. It is a strong flyer. Many of the British individuals seen in the spring will have migrated from mainland Europe, while others were emerging from hibernation here. They can be seen in many parts of the world where the climate is reasonably temperate – North Africa, Europe, Asia, The Americas and even the Caribbean.

A Noisy Interloper

On April 30th, my notebook says, 'Oh no. A parakeet'. I had heard an ear-piercing squeak, and shooting fast across the park was a sharp-winged, sharp-tailed and sharp-sounding green member of the parrot family, the ring-necked parakeet. I suppose I should not complain about a new arrival in Britain doing well as other species decline, but I'm afraid I do not much like the ring-necked parakeet.

It comes from a broad swathe of dry terrain from West Africa to India. They first began nesting in the wild in southern England in the 1970s after captive birds escaped or were released, and they are now common across the whole of the south-east. They roost communally in large noisy flocks and their screeching can be heard in every London park. There have been calls for the birds to be culled to protect our native species that also nest in tree holes, notably starlings that the parakeets bully out of their chosen nest sites, but also great spotted woodpeckers and nuthatches.

The RSPB is monitoring parakeet nesting areas to try to establish how they might be affecting native birds. Culling them seems pretty drastic to me, but I hope these raucous interlopers don't find their way in any numbers to my local parks. Nonetheless, the ring-necked parakeet becomes bird number 77 on my list for the park.

May

'All things seem possible in May'.
(Edwin Way Teale, American naturalist and
writer: 'Optimism')

Magpie

The Merry Month

Many people say that May is their favourite month, with long sunny days, fresh green growth everywhere, an abundance of flowers, cascades of blossom, and joyous birdsong. The name 'May' was taken by the Romans from their deity Maia, the earth goddess of growing plants. Small mats of bluebells are in full flower under the trees; some are the paler Spanish blue bells that stand up tall, others are traditional native bluebells that are darker with flowers down one side of the stem so that they droop.

Across the park, the song thrushes are at full mega-decibel volume with their 'fresh-peeled voices', as Larkin described. How can one not feel optimistic in the merry month of May? The blackcaps, chiffchaffs and willow warblers are singing. They have been joined by at least five sedge warblers with their striped heads and a scratchy, chirruping song, often delivered while the bird clings to a reed stem by the fishing pools. More butterflies are emerging and the meadowsweet is starting to come into flower with its cascades of blooms like pale yellow candy-floss. The trees are covered in young green leaves and their blossoms are attracting the bees and pollinating flies.

Whitethroats

This year the month begins with beautiful Mayday weather – warm sunshine and a few white clouds drifting lazily. In the park, I soon discover that the whitethroats have arrived from Southern Africa. Three of them are rattling their scratchy little calls from the sedges beside the pools. I very much like the whitethroat. It is a feisty little bird, larger than the chiffchaff and willow warbler, with handsome plumage combining buff, peachy and streaky-brown with a grey head in the male and, unsurprisingly, a striking white throat that puffs out when the male is in full song. Whitethroats tend to avoid human company, so we are lucky to have some in the urban park. They usually hide in hedges and brambles, and scold you with a repeated rasping squawk if you or your dog get too close.

But the sunny weather is provoking them into wasting no time in searching for partners, and two of them are boldly displaying. They climb almost vertically from the top of the brambles, chattering and burbling, then, flicking their fanned tails and stretching their wings, they flutter down in an angled descent before disappearing into the tangled undergrowth.

A few days later, I hear another migrant warbler calling from a dense clump of bramble beside the south side of the river. The song has a short trickling start, not unlike a white-throat or a blackcap, but after a second or two it turns into a loud trill, rather like a nightingale. This is the unmistakable song of the lesser whitethroat, another long-distance traveller from sub-Saharan Africa and Asia. As the name implies, it is slightly smaller than the whitethroat, and to my mind prettier, with white underparts and a dark mask across its eyes. It is also less shy, and can allow quite close views if you are lucky enough to find one in the bushes.

For a long time it was assumed that the whitethroat and the lesser whitethroat were very close relatives, perhaps even sub-species. After all, they look pretty similar. But recent science, in particular DNA analysis, has revealed that the two whitethroats are not as closely related as one would think. Apparently they have evolved in a similar way from two branches of the bird order of *passerines* (perching birds) and the family of warblers, *(silviidae)*. They are distant cousins at best.

Bobbing

Walking back along the riverside path I see a medium-sized bird with a brown back and white underparts shooting low over the river. Stiffly bowed wings with pale bars show that it is a common sandpiper. It veers into the bank and starts probing along the river fringe, with its tail pumping and bobbing in a characteristic way, a very attractive bird and number 78 on my park list.

As its name implies, the common sandpiper is not a particularly rare bird, but it is unusual in certain places and at certain times. In summer this is a bird of the north-west of England, Scotland or Ireland. This individual was travelling west along the Warwickshire Avon, probably heading for the uplands of Wales, having set off from its winter ground in Southern Africa or Asia a couple of weeks before.

After probing the river fringe for a few minutes, the sandpiper is on its way, uttering a brief high-pitched trilling call as it disappears under the bridge and past the towering walls of the castle. Why sandpipers bob has been the subject of many studies, with no clear answer. Some theorise that it helps disturb insects or distracts predators or attracts a mate. I don't really think any of these make much sense. The bobbing begins as soon as the young leave the nest, and solitary birds on migration bob and pump their tails as though it is a compulsive habit. I think the best explanation is that it helps the bird balance as it picks its way over slippery stones at the water's fringe. The dipper, bird of rushing streams, also bobs. But I am far from convinced! It remains a bobbing mystery.

Battle of the Corvids

The chiffchaffs, dunnocks and whitethroats are all calling in the calm, warm morning. A pair of magpies is foraging in the long grass beside the fishing pools. Suddenly, two crows fly fast past me silently from behind, one almost touching me with its wing tip, and they launch a full attack on the magpies. A rolling, flapping, chattering fight ensues, with one of the magpies pinned down on the grass, and the pair of crows pecking at its head until it lies still.

I think I have just witnessed a murder. Now I know why the collective noun for the carrion crow is a 'murder of crows'. The two assailants leave the lifeless victim and chase the other chattering magpie into a hawthorn hedge. Suddenly the apparently slain magpie jumps to its feet, takes off and heads towards the safety of the bushes. The crows are after it, and one even grabs it in its foot before the magpie breaks free and dodges into the thicket. How it survived the hammering to the head by the heavy bills of the crows I don't know. But it seemed unscathed by the experience.

Crows and magpies are never the best of friends, competing for territory as they do, and are often heard sparring in the trees. This is the first time I have seen a full-on battle for supremacy. It was revealing; the larger crows relying on brute strength to destroy their rivals, the wily magpie feigning death to get away and fight another day.

The magpies will have hungry youngsters in their nests by now. They build their nests very early in the year, presumably so that they can steal the eggs of other birds to feed their own young. Magpies are notorious egg thieves. Their nest has a roof. It is a huge ball of twigs with an entrance to the side. It is also likely to be decorated inside with coloured material. Orange or blue baling string is quite popular, as are chocolate bar wrappers. Why? No one has come up with a plausible answer. Perhaps magpies just have an artistic temperament and like a splash of colour in the living space. There is very little evidence that magpies steal metallic object such as coins or jewellery. Nevertheless the magpie has a worldwide reputation for pinching bright things, so providing a perfect plot-line for Rossini when he wrote his melodramatic opera, 'La Gazza Ladra' – The Thieving Magpie. A French serving girl is accused of stealing a silver spoon. No one else had access to the room. She was tried, convicted and executed. Later the true culprit is revealed to be a magpie that had nipped in through an open window.

One for Sorrow

But the magpie's reputation for endless curiosity and occasional purloining is just a small part of this bird's place in folklore around the world. In Britain there is probably no other bird that is associated with superstition as much as our bold, and boldly-plumaged chattering corvid that inhabits most urban parks.

In pre-Christian times, it was an important symbolic bird associated with good luck. Historians say the Romans believed that the magpie was highly intelligent and a solver of puzzles. Actually it is. Recent experiments show that most of the corvids are very quick to learn, and can even be taught to mimic human speech. In North America, some Native American tribes believed that wearing a magpie feather was a sign of fearlessness and cleverness.

But the Christian church gave the magpie a much more negative reputation. Apparently it was the only bird not to go into mourning at Jesus' crucifixion – clearly shown by its jaunty black and white plumage and iridescent tail-feathers. In France it was said that 'evil nuns' were reincarnated as magpies. The stories got worse. It was the only bird not to enter Noah's Ark, surviving the flood by sitting on the roof chattering and complaining about the rain. And magpies were said to carry a drop of devil's blood on their tongues. This piece of church-lore is interesting. Magpies were also regarded as dangerous gossips.

In 1568, Pieter Bruegel the Elder painted 'The Magpie on the Gallows'. It is at first sight a merry bucolic scene with peasants dancing to bagpipes and one relieving himself in the woods. But in the very centre of the picture there is a magpie perched on a gibbet. As with many of his pictures, Bruegel was making a political point and disguising it so that he would not be hauled off by the Spanish Inquisition. Spain's occupation of the Netherlands to suppress the dangerous rise of Protestantism was brutal. Bruegel was taking a considerable risk, but in failing health he may have thought he had nothing to lose. His magpie clearly represents the gossips and snitches who would send people to their deaths. Already in medieval times the magpie had stopped being a good-fortune symbol, perhaps because, like the other members of the crow family, it was present on battlefields feasting off the corpses, and at the gallows. In many countries it became a portent of death. The magpie's plumage looks like it is wearing the executioner's black hood.

Certainly the Victorians were fearful of magpies. These birds were untrustworthy; thieves and vagabonds; bad birds that brought bad luck. However, as the popular rhyme shows, it was generally thought that seeing only a lone magpie would bring bad luck, perhaps because most magpies mate for life, so seeing one bird will mean that it is lonely and in mourning.

'One for sorrow, Two for joy, Three for a girl. Four for a boy, Five for silver, Six for gold, and Seven for a secret never to be told.'

There are many regional variations. My Yorkshire parents would say '*three for a letter*'. It seems that across the country people still observe the superstition of warding off bad luck on seeing a lone magpie, by saluting the bird, or doffing your hat, or saying 'Good morning, general' or 'Hello Jack'. There are regional variations of magpie superstitions. In parts of Scotland, seeing a single magpie by the window of a house means there will soon be a death. In Wales, seeing a magpie flying from right to left as you start a journey means you might not make your destination in one piece, and in Yorkshire, where magpies are associated with witchcraft, if you see a single bird you should make the sign of the cross to keep the evil spirits at bay.

At the edge of the park there is a commotion in a line of silver birches. The trees are full of chattering magpies. I count eighteen of them, bouncing about the upper branches, flirting their long tails and making quite a racket. It is a magpie moot of some sort. Perhaps they are agreeing territorial rights across the park. After a few minutes they disperse in pairs. Magpies are very clever birds, not to be underrated. Perhaps they were indeed having a parley to avoid any damaging border wars across the park.

Superior Swans

The mute swans have produced a brood of eight cygnets. They are being introduced to the big wide world by their mother, the pen, who is leading them along the centre of the river, with the proud cob not far behind with his wings arched. The cygnets are incredibly cute; little balls of grey feathers with dark beaks all scrambling to climb aboard their mother, where they ride with their heads poking out. Five have made it on to mother's feathery back; the others are close behind complaining with their peeping.

Unfortunately it is highly unlikely that they will all make it to adulthood. Like the ducklings, young swans are food for many predators. But they will be defended fiercely by the parents that hiss and flap at any dog coming near, and if necessary will strike with their wings and beaks. The mute swan is not entirely silent as its name suggests. It will utter a number of grunts, hisses and soft growls, particularly when dogs approach.

Swans are pretty special, not only because of their extraordinary elegant beauty, but because they seem to know they are superior creatures, proudly bossing their reach of the river. Certainly the 'Swan of Avon' was an epithet given to Shakespeare because he was a bit special with a nobility of intellect, and of course there are lots of swans on the river in Stratford upon Avon. The swan is also deeply embedded in folklore around the world. In Britain, there are many ancient Celtic legends involving the white swan, often with a maiden or a child put under a spell that turns them into a swan, a spell that can only be broken by the handsome prince of course; or in some stories the prince himself turns into a swan to be with his love. In Greek mythology, the top god Zeus takes the form of the swan to seduce the lovely Leda. All over the world, the swan is the centre of stories of love, seduction and magic.

In the UK, all 'unmarked swans in open water' are owned by the monarch, though in medieval times the king or queen could grant ownership of swans on a particular body of water to the landowner. In London the Vintners' Company and the Dyers' Company still exercise some ownership rights and they organise the annual swan-upping on the Thames, counting, grabbing and ringing the mute swans on the Thames, then saluting the monarch as 'the Seigneur of the Swans'. Here on the Warwickshire Avon, the swans are unmolested by such rituals.

I can understand why in former times the swans were protected by royal ownership. Hungry peasants would be able to feed a family for a week on a nice fat swan. The meat is said to be tasty, rather like duck, 'moist and succulent' according to one chef. So they were reserved for the royal table; serving swan was regarded as the height of hospitality. Swan poaching could be punished severely, and taking swans' eggs was a serious crime deserving jail.

Two days later in the park, the brood of eight cygnets is still intact, but after another ten days there are only three. Could the other five have been snatched by foxes or badgers at night? Perhaps the rapacious mink had slipped past the parents. The pike cannot be ruled out. There is said to be a big one in the pools beside the river. I'm glad to say that the three survivors grew to maturity during the summer and autumn months, and were duly driven off by their loving parents to find their own territory.

Sea Swallow

This is an uplifting sight to see. In mid-month, a tern is fishing over the pools with its flip-flip wingbeat and sharp wings and tail. This beautifully slim, silver-grey and white bird is the common tern, not so common in the centre of the country. Again and again, this individual plunges into the water from about twenty feet, trying to catch a small fish, but without success. It is an extremely smart bird, with a black cap, a blood-red bill, and a forked tail that gives it the nickname, the 'sea swallow'.

Terns are certainly seen most often around our sea coasts, nesting colonially in areas of sandy gravel, but some breed inland on islands on lakes and pools, and particularly at reclaimed gravel pits. The common tern is another long distant migrant, arriving in Britain in the spring at the end of a week-long trip from its wintering grounds in West Africa. I'm inclined to muse about what this bird in the centre of England was doing just a few days earlier. Perhaps it was flying buoyantly over a fishing boat in the bright blue ocean off Senegal, or swooping past the surf boats of Accra off the coast of Ghana. Now it is flipping and turning over a pool in a park in Warwickshire.

After many attempts it dives into the water and emerges with a wriggling minnow in its red bill, and juggles it in flight before swallowing it. Success at last, and immediately it wheels away up river.

Mayflies

The river banks and pool fringes have turned bright yellow. There are tall buttercups in the marshy ground, yellow flags growing among the reeds, escaped oil seed rape along the river bank, and rafts of yellow water lilies coming into flower along the edges of the Avon. The birds are in full song, with the Cetti's warbler starting to sound rather desperate, and the blackcaps, chiffchaffs and greenfinches calling as they collect food to feed their young.

In the latter part of May the weather is calm, sunny and warm. The fish are jumpin' with the occasional loud plop, and the mayflies are rising. The mayfly is one of our most extraordinary insects, part of an ancient order that includes dragonflies and damselflies called '*palaeoptera*'. They are found in all parts of the world except Antarctica. There are about three thousand different species of mayfly, forty-one of them found in Britain wherever there is clean, fresh water. Those rising from the slow running river Avon this warm May morning, looking gold against the sunlight, are probably the common mayfly that has a pale brown body and greenish veined wings. The house sparrows have spotted them too, and have gathered in the bushes beside the river, chattering excitedly. They fly out over the water to expertly catch the rising insects and head off to their nest sites in the roofs of nearby houses and garages before returning for more easy food.

The mayfly has a famously short life out of the water and has become a worldwide symbol of the brevity of life on earth and the inevitability of mortality, noted in ancient times by Aristotle and Pliny. The larvae live underwater as 'nymphs' for up to two years, shedding their casings several times as they grow. In springtime they make their way to the surface and undergo two more acts of metamorphosis, first to a 'dun' with wings but unable to fly, and then the full adult mayfly with its lacy wings and three-pronged trail. They mate over the water, with the female falling to the surface where she will drop her eggs and promptly die, and the male won't last much longer; most expire within 24 hours. Life is certainly short for the mayfly.

The larvae reaching the surface and the adults falling into the water are a bonanza for the river fish. That is why fly-fishermen use the popular 'dun' fly, crafted to look like a newly emerged mayfly, to tempt trout and salmon. So here I am, watching the rising of the mayflies that has probably happened for millions of years, and one can't help reflecting on the transience of life. The writer Bill Bryson in his popular science book, 'A Short History of Nearly Everything', used a visual image to illustrate the tiny span of time that Homo Sapiens (or Homo Erectus of you prefer) has ruled the roost on the third rock from the sun. If the time of the existence of planet earth is shown as the span of the two outreached arms of Michelangelo's 'Vitruvian Man', one stroke of a nail file would remove the period we have been here as a species.

And in our individual lifetimes, do we want to 'make a mark'? Do we want to be remembered well? The Anglo-Saxons thought glory, achievement and a prominent life meant that your *name* would live on after your body had packed up – an immortality of reputation. The closing lines of the Anglo-Saxon epic poem, Beowulf, praising him at his funeral, describe him glowingly as 'the most eager for fame'. It's surprising to me how strong to this day is that need to 'leave your mark', more than a thousand years later. For some, fame is the spur that drives them on. For others, gathering rosebuds while ye may is life's mission. Watching mayflies makes you think. As a former journalist I am bound to recall that the poet George Crabbe likened the newspaper to a mayfly – ephemeral, otherwise described as tomorrow's fish and chip paper.

June

'Behold, now, where the pageant of high June
Halts in the glowing noon!
The trailing shadows rest on plain and hill;
The bannered hosts are still,
While over forest crown and mountain head
The azure tent is spread'.
(Bliss Carman)

Swifts

Feeding Time

The 'azure tent ' of the calm June sky is not in evidence this year. The month begins cool for the time of year with plenty of rain. Once again, the climate is unpredictable. This is certainly not 'flaming June'. After being away for a few days, I'm pleased to be back in the park, looking and listening.

The stock doves are producing a gentle, rippling cooing from a willow leaning across the river with its fronds tickling the surface. They have nested in a cleft where the tree was pollarded. The Cetti's warbler is singing loudly, and I think there may be two individuals, having heard some chip-chipping further away from the singing male. The whitethroats and blackcaps are feeding young and the trees are full of young blue tits and great tits, fluttering, calling and running their parents ragged. The adults look like they have been though a washing machine as they desperately try to find enough caterpillars and aphids to satisfy the clamouring broods. In an oak tree there is even louder 'see-see-see' begging sound. It is a family of nuthatches, with the parents racing up and down the branches looking for insects in the bark, closely followed by at least three fledgelings. They are paler than the adults and flutter their wings as they call, 'feed me, feed me'.

Spectacular Swifts

A faint high-pitched screaming indicates that the swifts that arrived from Africa in May are pairing up. Looking up I can see the sharp sickle-winged outlines of six of them moving high and fast against the pale grey clouds. Swifts are charismatic and fascinating birds. They are not closely related to the similar-looking swallows and martins of the *hirundinidae* family that also live by catching insects in flight; they are a more ancient and exclusive order called *apodidae*. They are more closely related to hummingbirds; but the two species really couldn't be more different, with one able to fly at zero miles per hour, and the other living its life at top speed.

Our European Common Swift can cruise at around 70 mph. It is beautifully aerodynamic. Aircraft designers have studied its scythe-shaped wing structure. How about this for a useless fact? In its lifetime, a swift can cover enough distance to fly to the moon and back five times over. This is because this bird spends almost all its life in the sky, feeding, mating, gathering blown nest material, and even sleeping on the wing, having developed a technique whereby half the brain snoozes awhile and then it switches on again to give the other half of the brain a nap. It uses its tiny legs only for shuffling about in the nest cavity or occasionally clinging on to brickwork.

Swifts migrate to the UK from sub-Saharan Africa, arriving a little later than the swallows and martins, then normally have just one brood before heading back in August. So finding a suitable mate is a speed-dating experience with no time to lose. Researchers recording and slowing down the distinctive screaming have established that this is probably a pairing-up technique with a male bird starting the scream and a female joining in at a higher pitch.

In Britain the swift is finding it harder to succeed in the procreation game. It is a bird in serious decline as a breeding species on these islands. The RSPB reports a 57% reduction in our population of swifts since 1995. The reasons for this decline are not entirely understood, but most studies conclude that lack of nest sites is a major factor. Swifts nest in holes in buildings, under the eaves and in roof spaces. Modern buildings and those upgraded to prevent heat-loss don't afford these entry points. There are plenty of initiatives around the country trying to stop the decline of the swift and to encourage these fabulous birds. Swift boxes fixed under the eaves are being installed in some schools and new-builds, and enlightened builders are using 'swift bricks' that allow the birds entry to the roof space. The Royal Estate of the Duchy of Cornwall has installed swift boxes across its properties. Oxford has declared itself Britain's first 'Swift City'.

Historic Warwick with its Medieval, Elizabethan and Georgian buildings has a healthy population of swifts in the summer. As they pair up, they hurtle in screaming squadrons through the narrow streets, as though they are on a bombing run against the Death Star in 'Star Wars'. Why they don't brain themselves as they swerve and shoot into a likely nest cavity I don't know. How do they put the brakes on? In the park, they hunt each day for flies over the river and the fishing pools. Sometimes a swift will pass by at speed very close and you can hear the air being torn like a ripped piece of cloth. Swifts must live with that tearing sound all the time. At the pools I watch as several swifts take turns to have a drink, shooting over the water and gliding with wings held in a v-shape as they scoop up water from the surface. You would think they would choke.

The week beginning on midsummer's day has been designated by conservationists as 'swift awareness week', publicising the need for more swift nest-boxes or swift-bricks to be installed to stop the decline in Britain of this fantastic bird.

Watching the Rails

The coots are back and have paired up already. There are two pairs along this stretch of the river. Coots are very territorial and argumentative, so there are plenty of spats between their two nesting zones, with the birds sitting back in the water and smacking their rivals with their feet while making a bit of a screeching racket.

The coots have been absent from the park all winter. For a bird that doesn't look like a powerful flyer, the coot travels surprisingly long distances to escape harsh winter weather, normally migrating at night. In Britain, huge rafts of coots, perhaps a thousand together, will gather on reservoirs or large lakes with the resident birds joined by coots that move to Western Europe from Russia and Scandinavia during the winter, before dispersing to their breeding areas, some of them to urban parks.

Coots are members of the rail family of birds, wild hens if you like. They are sometimes confused with moorhens, their family members, which I find rather surprising; the two species look pretty different to me. The coot is uniformly charcoal-black with a distinctive white 'shield' on its forehead and a white bill. Its close relative, the moorhen, is smaller, with a white stripe separating its dark brown wings from a dark blue underside, though it can look black from a distance. The bill is a brilliant red with a yellow tip, and under the tail are two large 'reversing lights' – white patches that are constantly being flicked up and down. It's quite an exotically coloured bird to be seeing so frequently in our parks. Another difference is their feeding habits. The coot dives underwater to collect pieces of weed; the moorhen stays on the surface. And when out of the water you can see that the coot's enormous feet have lobes of skin along the toes, whereas the moorhen has no webbing at all; nonetheless it still manages to swim pretty well. So all-in-all they are not very difficult to tell apart.

At the Kingfisher Pools, two pairs of moorhens have two chicks apiece. One pair had nested in a tangle of dead branches that had fallen into the smallest pool; the other pair had built their nest in a clump of growing vegetation at the edge of the main pool, just two meters from a fishing platform, but beautifully concealed. Moorhen chicks look comical with bald heads showing a pink patch. They follow their parents around peeping pathetically and learning from them how to pick up small pieces of weed and insects on the surface of the water. But these newly hatched chicks are very vulnerable. Moorhens can lay up to twelve eggs. These youngsters seem to be the sole survivors after just a few days. But their chances have improved, with the parents more able to protect just a couple of chicks rather than an undisciplined rabble. And they have the chance to affect an unusual escape when threatened. If a predator approaches, the young may cling to the parents' bodies, and the adult birds will fly away to safety, carrying their offspring with them – literally hanging on for dear life.

Moorhens are not found on moors. The name comes from the Anglo-Saxon meaning of 'moor' as a marsh. I think they were also called 'mere hens' inhabiting small lakes. They are often referred to as 'water hens' in literature. There is another member of the rail family living in the park. This one is much less common and much more secretive. The water rail is one of the most difficult of all British birds to see. Smaller than the moorhen, with a red curved bill, it hides deep in the reeds and is incredibly shy. I haven't seen a water rail in the park, but I know they are there, lurking in the damp areas of tangled vegetation by the pools, because other members of the local natural history society have occasionally seen them, scuttling across the path and even flying across the river. And I've heard one, making strange grunting and squeaking noises deep in the reeds. It's good to know that these elusive birds are here in the centre of town.

Flood Dangers

The first half of June is turning out to be very wet and cold for the time of year. It's a big contrast with the previous year when

June was extremely hot, but in recent times the weather trends in Britain have become unpredictable. A dislocation of the jet stream is being blamed for this year's miserable midsummer. This extremely high airstream snakes round the northern hemisphere as the world spins, bringing weather systems from west to east. This June the jet stream has become lodged over the British Isles and a huge area of low pressure is sitting across the country bringing rain and wind from the Atlantic.

In the park there are no swallows or martins to be seen, no butterflies or dragonflies, and the birds are quieter, partly because they are too busy feeding their young to pause for a song. The blackbird is an exception; it seems he can't resist singing in the rain. The Avon is running very high and fast. It is not quite the great grey-green greasy Limpopo of Kipling's story, but it is clouded with silt and noisily lapping at the banks. The river has stayed within its watercourse, at least for now. The fishing pools are on a large stretch of low-lying land on the south bank, part of the flood plain of the Avon, and Warwick is built on a hill, so not many properties are threatened here. But towns not too far away have not been so fortunate.

To the east there is serious flooding in South Yorkshire and Lincolnshire, where the deep weather depression decided to make camp and dumped rain on the same saturated land for days on end. And to the west, towns on the River Severn in Gloucestershire have also been affected. The Warwickshire Avon is a tributary of the Severn, the longest river in Britain, joining it at Tewkesbury, where the Worcestershire Stour also feeds into the mighty Severn. At Apperley in Gloucestershire, the Severn has by far the greatest measured water flow in England and Wales and when there is persistent heavy rain across the Cambrian Mountains and the English Midlands, the Severn can simply overflow.

Our changing climate means it is not just a possibility or even a probability, but it is a certainty that we will have more Atlantic storms and periods of heavy rainfall. More than ten years of reduced public spending has left the Environment Agency that is in charge of flood defences woefully under-funded. On the banks of the Severn they have installed flood barriers, but they admit that some of them will only push the problem of a swollen river further downstream, and the barriers themselves might not cope with a major flood. I can't help thinking that there is a flooding disaster waiting to happen in the English midlands.

Fascinating Dragonflies

During the second half of June, the low pressure system moves away and the sun is seen for the first time in three weeks. It brings the park to life immediately. With all the birds feeding their clamouring young and calling to keep contact as they forage.

And suddenly it is dragonfly and damselfly time. The first to show around the fishing pools are the banded demoiselles, the male a bright electric blue with the unmistakable mascara-like smudges on its wings, and the female a bright metallic green with no wing patches. They settle quite frequently allowing a close look. More difficult to spot is the aptly named azure damselfly. It folds its wings along its body when it alights. There's one – a bright blue male with black bands down its body and a prominent 'tail light'. The delicate demoiselles will soon be joined by the larger and more ferocious hawkers.

I think dragonflies are very difficult to identify. They whizz past quickly and seldom settle long enough for you to have a good look. And even if you do get a decent view, trying to count the rings on the abdomen or memorising the black and yellow pattern is rather taxing. The large common hawker and southern hawker are tricky to tell apart – the males with blue bands on the body, the females with bands of yellow or green. But the similar-looking migrant hawker that appears in late summer and autumn is smaller and more agile, and the brown hawker is much easier to pick out. It is big and brown and has golden wings. But I tend to think that naming them doesn't really matter. Dragonflies are beautiful, mysterious and ancient creatures that represent high summer in all its glory. They are also rooted in mythology around the world.

In some cultures, dragonflies represent good luck and prosperity. No wonder they are a favourite motif in jewellery. Traditionally anglers used them as an indicator of good fishing grounds. Plenty of dragonflies meant there were plenty of fish around. If a dragonfly hovered near the fisherman, he took it as a good luck sign. In various spiritual pathways, the dragonfly acts as a messenger between the worlds. If a dragonfly lands on you, you'll hear good news from someone you care about and catching a dragonfly meant you'll marry within a year. I think you would have to be rather desperate for a wife or husband if you try to catch a dragonfly. These extraordinary insects have five eyes and all-round vision and zoom away if you get too close. Trying to identify them as they whirr past is always going to be a challenge.

Mimics

The sunnier weather has encouraged the starlings to sing from the top of the old sea scouts' hut where they nest under the corrugated roof. There are plenty of starlings in the park, foraging in small groups across the grass, and flying back to their nests with worms or leatherjackets, the larvae of the daddy-long-legs, that they find by probing with their sharp beaks. The starling's song is unusual in that it is a wide range of squeaks, rattles, chuckles, piercing whistles and mimicry. The starling is famous for reproducing the calls of other birds and other sounds. The British Library records that, 'There is a reliable report of a starling in a London suburb that cried like a baby, and other starlings imitated V-1 flying bombs in London towards the end of World War II. Less reliable, though certainly within the bounds of credibility, is the account of a starling that mimicked a referee's whistle and upset a football match'.

There is certainly one starling in the park that sits on its usual perch on the boat house producing very convincing impersonations of house sparrows, buzzards, pied wagtails and magpies. I feel like applauding when it finishes its impressions act and returns to the serious business of feeding its youngsters. When the young have fledged and can fend for themselves, they form huge flocks in the countryside. Starlings are famous for their murmurations in the autumn when these flocks make extraordinary shapes in the sky like locust swarms, before diving into a reed bed to roost. The patches of reeds in the Warwick park are far too small to attract a starling roost.

And the jaunty starlings are in decline for reasons that are far from clear. A startling starling fact is that between 1995 and 2016, Britain's breeding population of common starling crashed by 51 percent. The situation in England is even worse, with an 87 percent decline between 1967 and 2015. Starlings no longer nest in some parts of Wales and southern England.

These birds are strongly migratory. In the winter, many starlings you see in the UK will have travelled here to escape the winter freeze in countries to the north and east. In those places too, numbers are falling, and we have fewer starlings arriving each year. It means that the large winter gatherings we are used to seeing are becoming smaller. Are the days of starling murmurations numbered?

The RSPB Centre for Conservation Science is trying to discover just what is driving starling declines in the UK and seeking solutions to improve the prospects for one of our best-known birds. They have looked at the effects of pollution, and the resurgence of predators such as sparrowhawks. But so far, no one knows exactly what is causing the drastic decline in the numbers of breeding starlings. The humble starling is now a red-listed bird, meaning it is of high conservation concern.

Cloudwatching

The improved weather at the tail end of June has produced some interesting cloud formations. This morning there is a ripple effect high in the sky with static clouds that look like a wide, flat, beach, just after the tide has gone out, with rows of ripples in the sand. Further to the east are some even higher wisps of white. These are cirrus, named after the latin word for a curling lock of hair. They are composed of ice crystals hanging in the atmosphere up to eight miles above the earth's surface.

I count myself lucky to live in a country where the sky is so interesting. The skies over the British Isles are seldom the same one day to the next, with moist air blown in from the seas that surround us. We may envy those who live in sunny climes, such as Australia or the Mediterranean countries, where clear blue skies can be guaranteed for weeks on end, but after a while it must get monotonous. The cloudy skies over Britain are an amazing canopy of shapes, colours and movement, and whether you are in the country or the town, every time you look up, the cloudscape is unique. And it's free.

There is a super book called 'The Cloudspotter's Guide' by Gavin Pretor-Pinney. He describes all the different cloud types and formations, and once you've read it you will never again look up without appreciating the shapes above our heads. Pretor-Pinney says, "The clouds are nature's poetry, and the most egalitarian of her displays, since everyone has an equally fantastic view of them. Clouds are for dreamers, and their contemplation benefits the soul. Yet their beauty is so everyday as to be in danger of being overlooked."

His book teaches us to appreciate their different varieties – the stratus, cumulus, nimbostratus and Morning Glory to name only a few – and all their beauties and significances, both meteorological and cultural. We learn how Hindus believed the cumulus clouds were the spiritual cousins of elephants, how thermal air currents act on fair weather cumuli, and how to save a fortune in psychiatric bills by using the clouds as Rorschach images that reflect our state of mind as well as nature's moods.

That is an interesting observation. Rorschach pioneered the test to establish what is troubling us in our subconscious by describing the images we see in inkblots. Seeing things in the clouds must be a pastime as old as the hills, and how interesting that our local bard had Hamlet, while feigning madness (or was he feigning?), teasing Polonius by looking at the clouds.

> Hamlet: Do you see yonder cloud that's almost in shape of a camel?
> Polonius: By the mass, and 'tis like a camel, indeed.
> Hamlet: Methinks it is like a weasel.
> Polonius: It is backed like a weasel.
> Hamlet: Or like a whale?
> Polonius: Very like a whale."

On the English summer days when the sky is cloudless, I think we appreciate the expanse of blue more than those living elsewhere who can take it for granted day after day.

July

'Daughter of pastoral smells and sights
And sultry days and dewy nights
July resumes her yearly place...'.
(John Clare. The Shepherd's Calendar).

Painted Lady

Sultry July

This year there are plenty of glorious blue-sky days at the start of July, with a high pressure area over The British Isles sucking hot and humid air up from the Bay of Biscay. When the cumulus bubbles up during the afternoons it certainly can become sultry, as John Clare described, and if towering clouds of cumulonimbus gather, there can be the occasional rumble of summer thunder. The air by the river is full of dancing insects. The martins and swifts are doing a great job of reducing the number of mosquitos over the pools. Amazingly a swift can gather as many as a thousand small insects in its large gape before heading back to its nest of growing youngsters with a visibly bulging crop.

Butterflies

There are also plenty of coloured butterflies flitting over the long grass. They definitely lift the spirits, for example after listening to the inevitably depressing morning news on the radio. It is hardly surprising that these gorgeous insects have become such important motifs in our art and traditions from time immemorial. Cultures from all parts of the world associate the butterfly with the soul. Psyche, the Greek goddess of the soul is sometimes depicted with a butterfly's wings.

Where does this association come from? For many, it appears to relate to the butterfly's sense of transformation. It starts out as a caterpillar, wraps itself in a cocoon and then emerges as a beautiful butterfly. In some belief systems, the caterpillar represents a life span. The cocoon equates to death. And the butterfly becomes the soul, fluttering away to the next stage of spiritual development. And not surprisingly, in some cultures butterflies represent rebirth.

Whatever they have meant throughout history, it is a pleasure to see butterflies in the summer months and it is fun to try to identify them. In just an hour I see small whites, large whites, and a green-veined white, (the identification is in the name), plus hedge browns known as gatekeepers, bright orange small tortoiseshells, a dark brown ringlet and the larger red admiral. In the dappled shade of the copse, the speckled wood butterflies are hard to miss. Small and brown with cream spots, they twirl together through the shafts of sunlight.

Beside the fishing pools the small butterflies are less prominent. The common blue seems to disappear into the vegetation when it settles, before flashing its pale blue upper wings as it moves on. But what's that one, settling on some ivy? The underside of the wing is almost white with small black spots, and when it opens its wings, the upper side is bright blue and attractively bordered in white. This is undoubtedly the holly blue, quite common in southern counties but unusual in the midlands. Butterfly Conservation says it is gradually colonising the midlands and the north of England. There are yet more butterflies to come this July and August, as I am shortly to discover.

A Painted Lady Year

Walking back along the river, a quite large ginger and buff-coloured specimen with black and white wing tips flutters past, then settles briefly on a thistle. It's a painted lady, a beautiful species rarely seen in the park. During the following weeks I am to see many more, as Britain is treated to a 'painted lady year' when there is a huge influx of these attractive insects arriving on a southerly airstream from North Africa and Southern Europe.

It is a once in a decade phenomenon. The previous painted lady year was 2009, when an estimated 11 million arrived in the UK, some travelling as far as Scotland and Northern Ireland. This year promises to bring an even bigger influx after the southerly winds in June and July have sent the thermometers to record levels in much of Britain. July 25th is the hottest day ever recorded in the UK, with 38.7% celsius measured in Cambridge. It is extraordinary that such a seemingly fragile insects can travel such vast distances. The painted lady that flew past me will have crossed the channel and southern England flying up to 1km high and at speeds of 30 mph, aided by the warm tail-wind.

Despite this good year for butterflies, the Butterfly Conservation Society says there is a serious long-term decline in the British population of lepidoptera – butterflies and moths. Why? They say that the destruction and deterioration of habitats as a result of land-use changes, such as more intensive agriculture and changing woodland management, are considered as the main cause of the decline of the butterfly species that are particularly dependant on certain plants and habitats. Butterflies are extremely picky. For example, the orange tip caterpillars will eat only garlic mustard, lady's smock or the sweet rocket in your garden. Several species such as the red admiral and the comma like nettles, and the common blue prefers to lay its eggs on bird's-foot-trefoil. So the loss of wildflower meadows, hedgerows and flower-rich fringes of fields is regarded as the main reason for fewer butterflies.

Mammals Great and Small

Apart from the bold grey squirrels, and the rabbits that emerge at dusk to feed on the grassy areas near the pools, it is unusual to see any kind of mammal in the park. Many are largely nocturnal, and they are simply too wary of people to venture into an area largely surrounded by houses and roads. Unfortunately there is still no sign of otters, though they are spreading along England's rivers as the water quality improves and the fish return. Water voles are far too timid to come into town, but the foxes certainly do; their dark droppings are seen regularly around the pools in prominent positions on the paths where they have decided to mark their territory.

So I was delighted to glimpse a tiny mammal running out on to a wooden fishing platform, pausing to sniff around with a long and flexible nose, then leaping into the reeds. Unmistakably a shrew, and with jet black fur it had to be a water shrew. These little animals have to eat about three times their own body weight of food every day to survive, and they will starve if they go for several hours without a meal. So they are always on the move, seeking out worms, snails and beetles along the water's edge, and diving under the surface to find freshwater shrimps or caddis fly larvae. The water shrew is most unusual amongst mammals in possessing venomous saliva. A mild toxin secreted into the saliva in the mouth helps to stun the prey, presumably to stop it wriggling away.

Nearby on a muddy track leading through a blackthorn thicket, there are signs of another mammal that inhabits the park, this one somewhat larger than the shrew. There are tracks in the soft ground, quite difficult to spot because they are no more than an inch long. Two parallel lozenge-shapes mean they are the 'slots' of a muntjac deer. This is really no surprise. All across England and Wales and parts of Scotland, the muntjac seem to be everywhere, even munching the roses in suburban gardens. This small Asian deer, no bigger than a large dog, is thought to have first bred in the wild in the UK after escaping from the Woburn Abbey estate in 1925.

A century later muntjacs are probably the most numerous of the UK deer species. A couple of times I have seen one in the nearby Priory Park, staring directly at me from the thicket with large ears and humped ginger back, before silently stalking away, ignoring the passing buses. But like most other mammals they will be more active at night and during daylight hours tend to lie up in the deep brambles. As the summer wears on, I am to discover there are more mammals to surprise me in the park.

Jenny Wren

The wrens are still incredibly vocal, hammering out their trilling song while trembling with the effort, their short tails held erect. It is an amazingly powerful song for such a tiny bird. Jenny Wren has always been one of Britain's most popular birds. Its image was on the farthing – the smallest coin of the realm – until decimal currency confined the farthing to the collector's cabinet. The wren is our commonest breeding bird, found everywhere apart from the very tops of the Cairngorms. They thrive in any habitat, and with their bustling style, always on the move, often scolding with their loud whirring alarm calls or exploding into song, everyone is familiar with the wren.

Their high-energy is impressive and endearing. In the breeding season a male will build several round, feathery nests in nooks and crannies, and the female will choose the one she likes. If a couple more females move into these designer abodes, the polygamous male will be more than happy to father several families at the same time, dashing about helping to feed several broods.

But why Jenny Wren? In medieval times, and probably long before, many common birds acquired nicknames; Tom Tit, Willy Wagtail. The jackdaw was a 'daw', as mentioned by Shakespeare in Much Ado About Nothing, and was probably nicknamed Jack because of its characteristic call, 'jack-jack'.

But jack also means small, as in jack snipe; the jackdaw is the smallest of the crow family. Its relative the magpie was originally just a 'pie' – and still is in France – but here in country districts it became known as the maggot-pie, shortened to magpie, probably because it fed off dead sheep and other rotting matter. The whitethroat in some parts is still called Peggy Whitethroat, or Nettle Peggy. I don't know why.

Jenny is a name often used for the female or the wife of quite a few animals. And in European folk legend, the wren was the bustling and scolding wife of the robin. "The Robin and the Jenny-Wren are God Almighty's cock and hen." (Traditional rhyme). For some reason the Robin Redbreast, originally called simply the redbreast, became known as friendly 'Robin', whereas Jenny Wren lost her nickname.

The wren is surprisingly prevalent in folklore. Killing a wren or harassing its nest is associated with bad luck – broken bones, injury to cattle, even lightning strikes on homes! 'Wren Day', celebrated in parts of Ireland and some other European countries on St. Stephen's Day, 26th December, features a fake wren being paraded around town on a decorative pole. Before the 20th century, real birds were captured, held in nets and paraded on poles by the 'wren-boys' who dressed in elaborate straw costumes and begged for money to be used for a Wren Dance that evening. The origins of these midwinter rituals almost certainly go back to pre-Christian times, perhaps as part of the druid rituals to celebrate the killing of the old year so that the new year could come to life.

In Christian times the origin of this rather horrible ritual is the notion of revenge for the betrayal of Saint Stephen by a chattering wren when he was trying to hide from his enemies in a bush. The Boxing Day ceremonies still feature some traditional songs, but curiously the wren is still honoured as the king of birds despite his betrayal of Stephen's hiding place.

> *'The wren, the wren, the king of all birds.*
> *St. Stephens's Day he was caught in the furze.*
> *Although he is little, his honour is great,*
> *Rise up, kind sir, and give us a trate'.*

According to Greek legend, the wren became the 'king of birds' in a contest to see which could fly highest. The wren hid on the eagle's back, and when the soaring eagle tired, the wren flew out to claim the title. And in Roman literature the noisy wren can't keep a secret. The famous author of 'The Twelve Caesars', Gaius Suetonius Tranquillus, wrote that the assassination of Julius Caesar was foretold by a wren. According to Suetonius, "a little bird called the king-bird flew into the Hall of Pompey, with a sprig of laurel." This 'king-bird' was a wren, pursued by a large flock of birds who wanted it to keep its beak shut. It entered the Roman Senate carrying a symbolic wreath to alert Caesar to his imminent demise. Unfortunately, the bird was torn to pieces before it could warn Caesar. Well, at least Julius Caesar got to have a month of the year named after him for posterity, the month of July.

Small is Beautiful

As I walk through the copse by the river, there is a constant very high-pitched sip-sipping and peeping, and looking up, I discover a family of goldcrests, seeking out microscopic insects in the tree canopy, often hovering underneath the leaves to pick off aphids. Even smaller than the wren, this is Britain's tiniest bird. The goldcrest is delightful to watch, and it is confiding, so will allow close views – if you can find it. As we get older, we are less able to hear sounds in the upper register, and many people over the age of sixty can't hear the goldcrest. I am lucky – so far. My hearing is pretty sharp. I really hope it stays that way. Being able to hear all the birdsong is incredibly important to me.

On this cloudy and rather sticky July day, the goldcrests are feeding four young that are chasing the adults through the branches, peeping their demands for spiders and aphids. The fledglings will have emerged from incredibly tiny eggs in a delicate nest made of moss and cobwebs, slung under a branch of a conifer. When the parents hang upside down to inspect the underside of the chestnut leaves, they show the bright gold stripe on the top of their heads – not a crest at all really. With their greenish plumage and fine wing-bars, they are terrific little birds to watch at close range.

Charms

Here in the urban park is another terrific species of bird - possibly the most colourful of all common British birds and also steeped in folklore. Beside the brook there is a stand of teasels, the dry deadheads much loved by flower-arrangers. They are also much loved by goldfinches. All summer they have been twittering and tinkling in the trees, and every now and then a group will drop on to the teasel heads to use their long bills to extract the seeds. It is normally only the males that raid the teasel heads; their bills are very slightly longer than those of the females, and that couple of millimetres is just enough for them to reach the seeds without having their eyes poked by the sharp spines. In Anglo-Saxon times, goldfinches were known as '*thisteltuige*' or 'thistle-tweaker', because of their fondness for thistles, teasels and knapweeds. The name teasel comes from the Old English 'taesl' meaning 'to tease'. The dried heads were used in the textile industry to raise the nap on woollen cloth.

Goldfinches are doing well in the British Isles. It seems they appreciate the warming weather. They have entered the top ten of birds seen in the annual RSPB Great Garden Birdwatch, probably because they are drawn to the seeds in bird feeders, and also because of milder winters. If the weather turns very cold, the goldfinch simply flies south. A large proportion of the UK population might migrate to Spain in the winter depending on the severity of the weather.

The collective noun for goldfinches is a 'charm'. They have certainly charmed people for generations. They are extremely colourful – not gold as the name might imply. That name comes from the bright gold wing bar that is easily seen in flight. The head is red, black and white and the plumage creamy buff. It is a dazzling little bird and also a compulsive songster, rattling out its cheery, tinkling refrains from dawn to dusk. So it is no surprise that the goldfinch was a favourite cage bird in Europe until the long-distance trading ships brought home canaries and budgerigars. As recently as Victorian times in the UK, many thousands of goldfinches were trapped and sold as cage birds, and one of the first campaigns of the Society for the Protection of Birds was directed against this trade.

This pretty finch also has a place in Christian iconography and folklore. The blood red face colour was said to have been caused by the bird being smeared with Christ's blood when it was trying to remove his crown of thorns, and with its association with thorny teasels and thistles, the goldfinch has become a symbol of the crucifixion. It appears in many religious artworks, particularly paintings of the Madonna and Child, or Christ with John the Baptist, with the bird representing the foreknowledge of the Crucifixion. Perhaps the most famous is Raphael's 'Madonna of the Goldfinch' in which John the Baptist offers the goldfinch to Christ as a warning of The Passion to come. So in church lore, the pretty goldfinch symbolises forbearance and endurance, and has also come to represent the Resurrection. Certainly in folklore it is regarded as a bird of optimism and good fortune. If the first bird you see on Valentine's Day is a goldfinch, it means your spouse will be rich.

By the end of July, the park becomes notably quieter. There is less birdsong. Even the vocal goldfinches have gone quiet. It must be difficult to be chirpy when your feathers are falling out.

August

'Silence again. The glorious symphony
Hath need of pause and interval of peace.
Some subtle signal bids all sweet sounds cease,
Save hum of insects' aimless industry'.
(Helen Hunt Jackson)

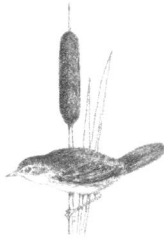

Cetti's Warbler

The Moult

As the nineteenth century poet Helen Hunt Jackson described, in August the birds fall silent giving way to the hum of insects. At first sight, nothing much seems to be happening in the countryside and parks in August. The birds are exhausted after feeding their demanding fledglings, and they are moulting.

The annual moult affects all birds, some quite significantly as their worn out feathers are replaced by the new, stronger ones. This can take many weeks, and a lot of our familiar birds seem to disappear in August and September. It is almost as though they are ashamed of their shabby appearance. They skulk in the undergrowth and stay silent, partly because their flight isn't as strong with worn or missing feathers so they are more vulnerable, but also because they don't need to impress a mate or compete for nesting territory. The blackbirds in particular, so confident and prominent for most of the year, seem to disappear entirely. The remaining twittering sounds in the trees and hedges tend to come from the young birds that are learning to fend for themselves; but they too keep a low profile being relatively easy prey for predators.

Walking beside the Avon I can see that the mallard drakes have been transformed. Most waterfowl, including swans, become flightless for a while during the moult and they must feel pretty vulnerable. The mallard drakes lose all their brightly coloured feathers first, and become almost indistinguishable from the brown ducks. It's called 'going into eclipse'. By the late autumn they will have spanking fresh new plumage, warm for the winter, and ready to dazzle the females in the next breeding season.

For birds that migrate, the moult is a particular challenge. They don't want to be embarking on an epic journey with worn out or missing feathers, and the process of growing new feathers is energy-sapping. Different species handle this in different ways. Small migrants such as chiffchaffs and willow warblers moult relatively quickly and they start losing feathers quite early, in July. So by the time they head south for the winter they have brand new flight feathers. Larger birds tend to moult after they have made the trip. Swifts are known to moult after they have arrived in their winter quarters. Buzzards and other birds of prey need to be able to fly well all year round, so they replace their feathers gradually over many months after the breeding season.

Therefore in August the park by the river has lost some of its vibrancy. The leaves seem to be dusty and the birds that are visible are not at their best. They are wary, because they are off colour in more ways than one. They are not as agile in flight and are lacking in energy.

Sparrowhawk Attack

Nature is never dull. Despite first impressions, there's plenty to observe and excite in August. I am walking along the impressive avenue of copper beeches when there is a sudden screeching and crashing overhead as starlings and pigeons rocket out of the trees. Something has zapped into the branches. Sure enough, a sparrowhawk drops out of the tree immediately in front of me and tries to make off with a flapping young starling clasped in one foot. A crow is in hot pursuit right behind it. I assume it is the usual mobbing of raptors by crows, but no. The harassed sparrowhawk drops the starling which flutters down to the grass, and as the predator makes off across the park, the crow wheels round and starts attacking the young bird, soon despatching it and immediately starts to eat it. This was not the usual driving away of a bird of prey, it was an opportunist theft.

Starlings are a favourite food of sparrowhawks at this time of year. The young birds not long out of the nest are naïve and plentiful. But all birds seem to hate sparrowhawks. The small male will take any songbirds; the larger females will regularly take wood pigeons. In the bird world, no one feels safe from the scary, yellow-eyed, dashing hunter. So when a few days later I hear a hysterical blackbird, panicking wrens alarm-calling and pigeons clattering out of the bushes near the fishing pools, I know they have been spooked by a sparrowhawk. And there it goes, a small male with a slate blue back and a rust-coloured chest, flipping away across the river having failed to make a kill this time.

Another Surprise

On the same morning that I witnessed the sparrowhawk having its breakfast pinched by a crow, another first for the park is clocked up, though rather briefly. I am standing by the Kingfisher Pools listening to the chiffchaffs that have now stopped calling their name and have resorted to a repeated 'phooeet' contact call. In the distance upriver I can hear some Canada Geese arriving with their loud honking. A squadron of a dozen birds, all braying, skims over the trees and splashes down noisily in the main pool.

The commotion flushes a brown, thrush-sized bird from the reeds beside the pool. It darts away jinking in flight and a glimpse of a bright white rump tells me it is a green sandpiper. It climbs swiftly and disappears over the nearby road to Leamington. This lovely, elegant little bird is an unusual one to see in a midlands town. The green sandpiper breeds in the far north, unusually for a wader nesting in trees, then migrates long distances to southern Europe. This one may well have been resting on its trip to Italy before it was disturbed by those noisy geese.

August was named by the Romans in honour of Caesar Augustus, who hugely expanded the Roman Empire and, according to St. Luke's Gospel, was the reason Mary and Joseph travelled to Bethlehem, when 'the whole world', (that's the Roman Empire), was to be 'taxed', (or registered in a census). I have also been cataloguing, counting and listing the different species I've managed to identify in St. Nicholas Park. With the great white egret in March, the parakeet and the black kite in April, and now the green sandpiper, my bird count for the park rises to 78.

Bulrushes

In the second half of August it gets hotter with sunny skies and just a light breeze. The dragonflies are active, with the hawkers patrolling alongside the thickets, and they have been joined by several common darters, small, red dragonflies that settle on the path in front of you, then flit a little further along when you get to within three of four yards. And what's that perched on one of the reeds? It's a small damselfly with a green body, a blueish thorax and black eyes. It must be a female azure damselfly. Good to see close up. The name comes from the bright blue male.

The bulrushes at the edges of the pools are attracting the attention of a pair of reed buntings. The seed heads like large Havana cigars are starting to disintegrate into a cotton-like fluff that drifts away on the breeze, mingling in the air with the floating seeds of the rose bay willow herb and thistledown. The buntings are clinging to the tops of the rushes, nibbling away at the tiny seeds, (they are less than a quarter of a millimetre long), attached to each individual hair, helping to send clouds of down across the water. Bulrushes, also known as reedmace, are interesting plants. They are 'monoecious', meaning they have flowers of both sexes. The small male flowers form a narrow spike at the top of the vertical stem, and wither away after shedding their pollen. Below, large numbers of tiny female flowers form the dense rich-brown sausage that we know so well on the bulrush. When the seeds inside are ripe, the flower heads disintegrate and the wind, or birds, distribute the seeds.

There are plenty of other intriguing plants growing in the damp ground near the pools. Matching the bulrushes for height, the purple loosestrife and great hairy willowherb are splendid, their flowers almost magenta in the sunlight. Lower down there is a plant with bright yellow buttons for flowers. It is a tansy, unusual in this part of the world. Tansy was cultivated by the Ancient Greeks and has been grown as a medicinal herb across Europe for centuries. It has a pleasant smell but the leaves and flowers are toxic to most creatures, so it has been used to rid people of intestinal worms and as an insect repellent. For many years it was used in embalming, to keep those worms away, and a tansy wreath was often popped into a coffin. In the American colonial period, meat would be wrapped in tansy leaves to keep the insects off. But it's not a good idea to take more than a small dose of tansy medicine to kill off your intestinal worms; the toxins can cause convulsions and even brain damage.

Further along the bank is another plant that country people have used over the years. It's a straggling plant about three feet tall covered in sky blue flowers. This is chicory – a very nice plant to find and again not very common these days. Sometimes called the blue dandelion, for centuries chicory has been cultivated for its salad leaves, its 'chicons', (blanched buds), and famously for its roots which have been used to make, among other things, coffee substitute. In recent years, an extract from chicory root called inulin has been used in food manufacturing as a sweetener and a source of dietary fibre. There is more to these pretty flowers beside the Avon than meets the eye.

Invaders

Near the sea scouts' landing-stage there's another interesting-looking plant growing waist-high and sporting some exotic speckled and orange flowers shaped a bit like an old-fashioned policeman's helmet. It is orange balsam also known as jewelweed – an invasive species from North America becoming more common in the English midlands. It is also known in parts of the USA and Canada as 'spotted touch-me-not'. The native Americans discovered that the juice of the leaves is a useful remedy for poison ivy, but unfortunately some people are very sensitive to orange balsam and will get a severe rash if they come into contact with it. I didn't touch it.

Across the river on the far side of the pools, there are patches of another invader, now well known across Britain and heartily disliked. The Himalayan balsam has attractive pink blooms and was first introduced into the UK as a garden plant in 1839. But it soon escaped and spread along watercourses and across damp areas, smothering native plants as it went. This tall invader is fast growing and spreads quickly. Its ripe seed pods explode, firing the seeds into the water where they are carried downstream to create riverbanks banks fringed with aliens. Its flowers are particularly attractive to pollinators, and rising above native species, the balsam tends to draw in the bees and flies, and the traditional British plants miss out.

In the park in Warwick, groups of volunteers uproot the plants each year to protect the diversity of native species. This has to be done before the seed pods explode, and also the uprooted plants have to be treated carefully. If left in a pile beside the water, they can manage to root themselves or still spill their seeds. Triffids come to mind. So they must be left on a plastic sheet so that they dry out and die before being composted. Across southern England, the Himalayan balsam is flourishing, clogging water courses and stifling native plants. It is up to volunteers and groups like the network of Wildlife Trusts to spend time and energy to keep this invasive species at bay.

Amazing Flying Mammals

When I had been noting in my diary how difficult it is to see mammals in the urban park because many are nocturnal, I had been thinking of deer, badgers, mink and otters. But of course the other group of mammals that we rarely see are the bats. In August, with so many flying insects about, the park must be a playground for bats. But which species and how many? Fortunately I know someone who will have the answers.

Tricia Scott is the Chair of the Warwickshire Bat Group and an expert in these flying mammals. As one of a network of carers co-ordinated by the Bat Conservation Trust, she even rescues and rehabilitates bats. At any time of the year, but particularly between April and September, she may be sharing her home with injured or exhausted bats, or young ones that have fallen out of a roost, caring for them until they can be released. So I asked Tricia if she would take me for a bat walk in St. Nicholas Park one calm August evening, to see what was there after dark. She was only too pleased to share her enthusiasm for bats. It proved to be one of the best wildlife experiences I have enjoyed for many years.

By now in the last week of August, the weather was hot, record-breaking in fact. It was the hottest late August bank holiday on record, 33° celsius in Warwick during the day and completely calm; perfect bat weather as it turned out. Tricia had with her a sophisticated bat detector, an ultrasonic microphone plugged into an iPad which converts the high-pitched calls of the bats at a ratio of 1 to 10 into the range humans are capable of hearing. The iPad also displays the calls on the screen as a sonogram, so that we could 'see' the different sounds and identify them. Bats use sound to navigate and catch insects, keeping up a constant stream of rattling and chirruping sounds.

Quite quickly as we walked in the dusk along the riverside, Tricia was able to identify four species. At the copse there were lots of *soprano pipistrelles* swooping round the trees, making the detector rattle with sound. They were clearly visible against the fading evening sky and swooping low over our heads. I even ducked on a couple of occasions as a bat seemed to be heading straight for me. The bat detector also revealed some *common pipistrelles* with lower pitched calls.

Then further along the river the slower and deeper clicking revealed a *noctule bat* passing by. They fly very fast – 30 mph apparently. Our largest native bat, the noctule has longer wings to power its fast flight. But where were the rarer *Daubenton's bats* that only feed over water? The clicking of the detector revealed that one or two were present over the slow moving Avon, but in the fading light they were impossible to see. Tricia explained that Daubenton's fly low over the water, sometimes picking insects off the surface, "like little hovercraft". It was really fascinating to know there are so many bats in the park, unnoticed by the late night dog walkers, and we had managed to find 4 of the UK's 17 native species.

Are they blind as a bat? Tricia explained that bats aren't blind; in low light they can navigate by sight, but like us they can't see when it's dark. To catch insects at night they have evolved an incredibly sophisticated system of echo-location, bouncing sound off their prey and the surroundings to guide them to their food. In the same way that sunlight reflects off surfaces and our retina makes a detailed picture, their series of rapid clicks rebounds and gives them a very clear picture of their surroundings – in such detail that they can "see" and catch a fast-moving mosquito.

Night Sounds

There is even more to be learned from night walks with a sound detector. Set at the right frequency, these devices can pick up the sound of insects communicating after dark. And it is quite extraordinary to hear the cacophony of sound beyond our audible capabilities. Tuned in and amplified, and when pointed to an area of thick grass, the microphones relay to us a symphony of buzzing, chirruping and wizzing that is surprisingly loud, and shows that after dark the insect world likes to party. It's all about linking up and mating.

It was a great experience to hear the bats and the insects. And in the distance, just to make the atmosphere perfect, a tawny owl was hooting. The idea that they say 'Tu-whit, tu-whoo' isn't quite right. The whoo is the male and the tu-whit is the female, or rather she sounds to me rather more like 'ke-weck'. The woo-hoo of the male tawny owl has become a cliché on film and TV soundtracks for night-time scenes. It certainly is a haunting sound. The bird emits an initial 'woo', then waits four or five seconds, before concluding with a tremulous 'woohooh'. It is said the period of the delay is ten heartbeats of a rabbit.

Breeding Cetti's

The end of August remains hot with clear skies and hardly a breath of wind. The song birds are fairly quiet, most still moulting. Five late swifts are fuelling up over the ponds before heading south. There is an adult peregrine preening on the tower of St. Mary's Church, but disappointingly the falcons clearly haven't nested there. It seems they have been using the ledge on the tower as a dining table and a place to rest between hunting sorties.

But one unusual bird has indeed bred successfully in Warwick. A loud "chip" in the reeds beside the pools in the park means a Cetti's warbler is hiding in the undergrowth. Then immediately a second bird seems to answer the first with a sharp trill from another clump of reeds ten yards away. So there are definitely two. The male's urgent singing throughout the spring was not in vain. Then the first bird appears on a reed, gives its sharp "chip" and flies away into the hawthorns. It is a juvenile – a bit smaller, with a slightly shorter tail and a little more grey than rich brown. So Cetti's warblers have managed to produce young in the park, probably for the first time.

September

'September comes, and Nature falls
silent. She has the work of harvest
to perform. You're on your own now
to map the sun, to touch a green leaf
turning yellow, to hold a golden moment
before it fades into winter light.

The days grow shorter, the nights
colder. Oh, Misery! you think.
And yet...'
(British poet, Daniel Brick)

Cormorant

Indian Summer

This September continues where August left off, with very warm and calm weather. After a few days it becomes breezy and cooler but then becomes flat calm again. It is turning into an 'Indian Summer'. The origin of that expression, meaning an unseasonably warm and dry spell in the autumn, is not very well known. It was used in the USA referring to Native Americans well before it was recorded in the UK in the 1950s. So the phrase does not refer to the weather in colonial India.

Insects abound, with plenty of wasps foraging among the brambles, and the occasional brown and yellow hornet zooming past with a rather scary loud drone. The damselflies must be wary. The insect-eating hornet will be happy to grab a damselfly to take back to its papery nest. There are larger dragonflies in abundance now, with hawkers buzzing around the pools and patrolling the riverbanks with clattering wings. And there are still plenty of butterflies, notably the brightly patterned painted ladies in this year of an invasion by this migrant species.

The Black Death

There are ripples in the centre of the main pool and after a few seconds a black, snaky shape emerges – the head and neck of a cormorant. This voracious fish-eater was once confined to sea coasts in continental Europe, with numbers kept down by human persecution and pesticide pollution in river estuaries. But in the mid 20th century they expanded their breeding range all around the coasts of the British Isles, and in recent years have begun to breed inland. In 1981 the first inland breeding colony appeared in Essex with the birds nesting in trees, and they continue to expand their range across the country. Others move inland from the coast during the winter. Perhaps winter storms make sea fishing more difficult. This individual might have left its breeding grounds on the Atlantic coast and followed the course of the River Severn into the midlands for a winter of less challenging fishing.

The water in the park pools is flat and the cormorant dives again, leaving slowly expanding concentric circles on the surface and staying underwater for about 30 seconds before reappearing 50 yards away. It must have good lungs. In China and Japan, cormorants are known for being trained by fishermen to do the fishing for them, sometimes diving for up to two minutes at a time. They are deployed from small boats with a line tied round the bird's throat, tight enough to prevent it swallowing a large fish. The fishermen are able to retrieve the fish simply by forcing open the cormorants' mouths; apparently this engages the regurgitation reflex and the fish pops out.

But here in Warwick there is no partnership between fishermen and cormorants. Anglers see them as very unwelcome rivals, and have nicknamed them, 'The Black Death', claiming they can clear a fishing area of large fish in just a few weeks. Owners of inland fish farms and fisheries say they are suffering high losses due to these birds, and for more than twenty years, angling associations have been lobbying the government to allow cormorants to be killed under a general licence, similar to those issued for the control of crows and wood pigeons. They complain that current licences allow holders to shoot only a handful of birds a year, and say that is not nearly enough when the number of over-wintering cormorants in Britain has risen from around 2,000 in the early 1980s to about 25,000 now. In 2004, the government announced that up to 3,000 birds could be killed each year in areas that were suffering severe damage to fish stocks, enraging the RSPB who had not been consulted, and claiming there are ways of scaring off cormorants rather than shooting them. The arguments continue. The bird remains protected under the Wildlife and Countryside Act as are all wild species in Britain, but licenses for shooting limited numbers are granted each winter.

Disturbed by some walkers with dogs, the cormorant I have been watching takes off, noisily pattering the water with its feet as it gains speed, and settles in a tall ash tree overhanging the river, where it spreads out its wings to dry in its iconic dragon-like shape. The structure of their feathers does not retain air droplets. While fishing, their wings soon become waterlogged, to allow the birds to dive deep and move fast underwater. So in order to be able to fly strongly and not catch a cold, they must hang their wings out to dry after a spell of fishing. As the winter progresses, a cormorant silhouetted against the sky on the ash tree, wings outstretched, can be seen quite regularly.

Anglers' Tales

The following weekend the local angling clubs have organised a fishing competition in the park at the Kingfisher Pools and along the Avon. The anglers arrive at their allocated stakes with the most extraordinary amount of gear, like roadies setting up a Rolling Stones concert. There are comfy chairs and large umbrellas to be unfolded, keep-nets in keep-net stands, a collection of rods and some enormously long extensions that mean they can drop the float exactly where they want in the middle of the river or pool. There are boxes of floats and hooks and bait in plastic containers, ranging from old-fashioned maggots to artificial lures, and ground bait that can be thrown into the water in handfuls to attract ground-feeding fish like gudgeon and roach. And of course a box of sarnies and a few cans.

I have never fallen prey to the lure of angling. But millions have. Fishing is the biggest participation sport in the UK, and the midlands is its heartland. Brummies are characterised as always leaving the wife and kids on a Sunday to go fishing. One of my favourite Brummie jokes is this conversation on a canal bank:

> "Have you caught anything?"
> "Yeah. I caught a whale".
> "Where is it?"
> "I threw it back".
> "Woy?"
> "Didn'ave no spokes in it".

I watch Clive setting up his equipment at stake number 13. He's happy to talk before the whistle blows for the start of the competition. I am not daft enough to talk to an angler while he is competing. I ask him if he is superstitious. He says he isn't. Then says, "I'll do alright here, touch wood". I ask him what he's hoping to catch. "There are lots of fish in here: roach, bream, tench, perch, and a few pike. There are some carp that get into the pools from the river by way of a culvert. If you catch a carp, you're supposed to put it back in the river". He doesn't seem to mind the predatory pike known to be in the pools. "They tend to pick off the sick and slow fish. There's plenty for all". And what about the cormorants? "They are a menace. But so long as there are only one or two... If a cormorant arrives while we're fishing, we yell at it and throw stones and it soon scarpers." Herons? "Not a problem. They take a few small fish, but there's plenty for all and it's good to see them".

I reckon the wide appeal of coarse fishing combines a lot of attractions. Yes, it's a bit of peace and quiet from the family home and the working week – some slow time to enjoy nature and relax – the camaraderie and banter with fellow anglers – but I detect that it's more exciting than it seems. Watching the float, waiting for the bite, knowing more and more about the behaviour of the prey, and the satisfaction of using all your expertise to snare the big one is an almost primitive hunting instinct that I recognise as a birdwatcher. If I want to see a rare bird, I get to know exactly how it behaves, where it might be and what it sounds like. A birder will say, 'I had a redstart today', in the way that an angler will 'have' a nice roach. I don't know if Clive won his competition from stake number 13. If he did he would have been genuinely elated.

The Dabchick

A few days later, there is another fish-eating bird working its way along the far bank of the river. It is like a ball of fluff floating on the surface but every now and then skittering a few yards along the surface, especially when the ducks get too close for comfort. This is the very attractive little grebe, also known in country areas as the dabchick. In the breeding season the adult has a bright chestnut-red throat and cheeks, and a bright white patch in front of the eye. It's like a dab of brilliant white paint; is that why little grebes were known as dabchicks? But this one has already developed its winter plumage of soft greys and browns.

The little grebe is a fast underwater swimmer. It has to be – to be able to catch the small fish that are its staple food. Just before it dives, I can see its powder-puff feathers flatten down suddenly, so that by the time it is under the surface, the air has been expelled from its soft plumage and it is a lean underwater machine. This one is doing quite well. Every fifth or sixth dive, it surfaces with a wriggling silver fish, which quickly disappears.

It is a lovely bird to see on the river in the park. The little grebe is quite a shy bird, often hiding in the reeds. But it sometimes betrays its presence by having a noisy argument with a neighbour, perhaps with a moorhen twice its size, and makes a loud noise like high-pitched whinnying of a horse, or to my mind a hysterical laugh.

In the 19th century, little grebes and in particular their relatives the great crested grebes had a very hard time at the hands of dedicated followers of fashion, and these bird species have two remarkable campaigning women to thank for their present healthy populations in the UK. In Victorian times, feathers were the height of fashion, particularly in lavishly decorated women's hats. There was a thriving trade importing exotic bird feathers from around the world, and here at home, herons, ospreys, kingfishers, owls and grebes were shot for their feathers. Grebes were particularly at risk because their soft feathers were also used as a substitute for fur on coats; the great crested grebe came close to extinction.

In Victorian England, women couldn't vote or own property, but this didn't stop a group of passionate individuals challenging the status quo and standing up for a cause they believed in. It began in Manchester, where in 1889 Emily Williamson established 'The Plumage League' from her home to oppose the use of feathers in fashionable hats. The first members were all women. Despite being mocked in magazine articles – written by men of course – the group quickly gained popularity and two years later Williamson joined forces with fellow campaigner Eliza Phillips, head of 'The Fur and Feather League' in Croydon, to form 'The Society for the Protection of Birds'. Just 15 years later, in 1904, the organisation was granted a royal charter. Now the RSPB has over a million members and is the largest wildlife conservation charity in Europe. Hats off to Emily Williamson and Eliza Phillips!

On the Move

It is scarcely noticeable but large numbers of birds are now on the move as the mass migration of autumn gets underway. The black-headed gulls are starting to trickle back to the park from their breeding areas, mostly adults but with some young birds that can be identified by the brown feathers on the back of their wings and a black band on the end of their tails. On 5th there are three swallows fluttering and skimming over the pools – one adult and three juveniles with much shorter tail-streamers than the parent bird. They will be feeding up before heading south on their long journey through western France, across the Pyrenees, down eastern Spain into Morocco, across the Sahara and finally arriving at their winter territory, perhaps a reed bed in Namibia or the open plains of a game reserve in South Africa.

A week later there is a lone swift flying high and fast over the park. It looks like it knows it is late and is hurrying to catch up. Most of the swifts that nested in Warwick left in August. But there are still plenty of warblers around, mainly whitethroats and chiffchaffs, making the most of the insect-rich water margins in the unseasonably warm sunshine. As soon as the weather turns chilly, they too will be on their way south. A thin 'tsi-tsi' overhead makes me look up, and a meadow pipit is passing over. It's a common bird in upland areas and open countryside, but far from common in the town. In autumn the meadow pipits abandon the moors and hillsides, moving to lowland areas where the winter weather will be less harsh, and some migrate to continental Europe. This lone bird might be on its way to a winter break in Spain.

Jizz

The characteristic high pitched call identifies the meadow pipit even if it is too far off to see the heavily streaked plumage, but its flight helps identification too. It is slightly bouncing, but is not as fast as a goldfinch that also has a bounding flight pattern. This kind of characteristic movement or general appearance is known to naturalists by the word 'jizz'.

The wildlife writer Sean Dooley described jizz as "the indefinable quality of a particular species, the 'vibe' it gives off". Birdwatchers are very familiar with the word. For me it is a combination of movement, posture and shape that can tell you at least the family of the bird and quite possibly the species, even at a distant glimpse. There has been much speculation about where the word comes from. One theory is that it comes from the World War II RAF acronym GISS for 'General Impression of Size and Shape' (of an aircraft). But an article in British Birds magazine by the naturalist Jeremy Greenwood and his brother Julian, seems to disprove that, first because there is no evidence at all that the RAF used such an expression, and secondly because jizz is in print as the general appearance of a bird in 1921.

Research by the Greenwoods had indicated that the word was introduced to ornithology by the respected Cheshire naturalist T.A. (Thomas) Coward in his Country Diary column for the Manchester Guardian in December 1921. The article was subsequently used in his 1922 book, 'Bird Haunts and Nature Memories', and attributed it to a 'west-coast Irishman'. He explained, "If we are walking on the road and see, far ahead, someone whom we recognise although we can neither distinguish features nor particular clothes, we may be certain that we are not mistaken; there is something in the carriage, the walk, the general appearance which is familiar; it is, in fact, the individual's jizz". Another theory is that jizz is a contraction of 'just is'. Perhaps that is why the 'west-coast Irishman' used the word. Whatever the origin, it is a useful word to describe the general look of a bird.

A good example of the way we can identify a distant bird by its jizz is to compare two common birds of parks and gardens, the dunnock and the robin. They are the same size, generally brown, and glimpsed from distance are classic SBBs. But the dunnock is usually skulking whereas the robin is confident and visible. The dunnock is constantly on the move, often flicking its wings. The robin when it lands is immediately still for a moment, as though posing, and it stands erect. It might bob – the dunnock doesn't do that – then the robin will flit to another position and pose again. Even against the light and many yards away, you can tell which is which by their jizz.

October

'Season of mists and mellow fruitfulness,
Close bosom friend of the maturing sun;
Conspiring with him how to load and bless
With fruit the vines that round the thatch eaves run'.
(John Keats)

Redwing

Wind, Rain and Thunder

Keats' famous *Ode to Autumn* is an idealised description of the season of bounty with its 'mellow fruitfulness'. This year the end of September and the early part of October is far from mellow. Three days of torrential rain and high winds across England have brought local flooding – part of the Birmingham orbital motorway was closed for a while with water axle-high on the road surface – and there have been brilliant shards of lightening against the purple sky followed by cracks of thunder.

I have always loved electrical storms. As a boy, if woken by thunder, I would press my nose against the bedroom window watching for forked lightening jagging down, and would love the tearing sound of a thunderclap before it broke into a series of crashing rumbles. How could there be so much electricity in the sky?

To this day, the exact mechanisms in clouds that causes thunderstorms and the power of the electrical charges they can generate, are not fully understood, and thunder and lightning are the subject of many studies. In 2019, scientists in Mumbai used a telescope designed to study the cosmic rays that bombard our outer atmosphere to look into a large thunderstorm and they concluded it contained 1.3 billion volts of potential energy. That's ten times more than had been previously estimated. I think you have to have several degrees in physics to be able to explain thunderstorms. I haven't and I can't. But I know that the power generated, just from interactions between ice and water, is literally awesome. It's hardly surprising that every civilisation in the world feared and admired their thunder gods, whether it be Zeus for the Greeks, Indra in Hinduism or Thor for the Vikings. In North America, many native tribes worshipped the powerful Thunderbird who made the thunder by flapping his wings, and they carved him on their totems.

New Arrivals

In the park the River Avon is brown and swollen, submerging the fishing platforms and creeping over the banks across the flood plain. The centre of the river has a line of fast moving flotsam. A large tree branch threatens to torpedo an intrepid lone canoeist who quickly paddles out of the mainstream. On the railings opposite the duck-feeding area, a heron stands in the drizzle on the top rail with two cormorants perched evenly spaced a few feet away on either side. It looks like a finely dressed member of the aristocracy with two rather menacing bodyguards. There will be no fishing for them while the water is so muddied and turbulent.

But the winds generated by this deep low pressure system have brought in the winter visitors from the north and east. Some very high-pitched 'seeps' in the copse by the river indicate that the redwings have arrived. These early winter visitors from Scandinavia and Russia have braved the rough weather on their five hundred mile flight across the North Sea, most migrating at night, to be first to get to the berry-rich hedgerows in the British Isles. Some will not have made it, falling exhausted into the sea. Redwings are our smallest thrush, a little smaller than the song thrush, and they are very attractive if you get a close view, with a cream supercilium, that's the stripe above the eye, a heavily streaked breast and gingery red on the flanks. So it's not really a redwing, it's a redflank.

When I was a schoolboy, I first noticed them on the playing fields outside my classroom window and wondered what these small stripy-headed thrushes were. I looked them up in a book in the library – ah, redwings! I was intrigued to discover how far they had flown to be in Britain for the winter and watched them with more interest, until a stern voice from the front would shout, "Thompson! Stop daydreaming and pay attention!"

The redwings are accompanied on their migration by their larger relatives, the fieldfares, with their bright plumage of rich brown, slate grey and cream. They seem to be more wary than the redwings, favouring the open country rather than an urban park, but they can often be heard flying over uttering their characteristic 'chak-chak' call, an atmospheric wild sound that indicates that winter is on its way.

A Surprising Diving Duck

After a few days, the wet and windy weather moves away to be followed by warm sunshine and the river becomes more benign. Walking beside it, I glimpse a dark bird near the footbridge just before it dives below the surface. It wasn't a coot. I'm sure it was brown rather than black and it had a bulkier head. Possibly it was a tufted duck that also dives for food and is seen occasionally in the park. I wait, and not twenty yards away a duck bobs to the surface. It has a dark brown head with a pale collar, soft brown plumage and a flash of white on the wings. It is a female goldeneye. I can clearly see its bright yellow eye before it dives again, searching along the river bottom for crayfish, snails and underwater insects.

This is a bird of sea coasts and lakes, so a very unusual one to turn up in the middle of town. The goldeneye is another long distance traveller in the autumn. It breeds in tree-holes in the part of Europe known as Fennoscandia – northern Norway, Sweden, Finland, Karelia and the Kola Peninsular in Russia, an area covered in snow and ice in the winter. So in autumn the goldeneye will head south and west to spend the winter in more temperate waters. This one might be heading for the huge Blithfield Reservoir in Staffordshire, and then, after a rest, might head off to the Irish Sea. She seems undisturbed by this human standing nearby on the bank, and swims upstream under the footbridge, unnoticed by a group of dog walkers. It is a lovely bird to see so close to home and completely unexpectedly. The goldeneye becomes bird number 79 on my list for the park.

Halcyon Days

By mid-October the weather has become bright and very calm, with the pools like mirrors reflecting a line of white puffy clouds in the blue, and some late dragonflies fluttering around the fading reed beds. The irrepressible wrens are still trilling from the brambles where the blackbirds are picking off the blackberries and the dunnocks and great tits are feeding on the glossy black elderberries. Some people with plastic bags are gathering the deep purple sloes from the blackthorn hedge. I ask one couple how they make Warwickshire sloe gin, thinking I'll get an interesting traditional recipe. "We get some gin and put sloes in it. That's all". Obviously. I shouldn't have bothered to ask.

These periods of warm calm autumn weather are sometimes called 'halcyon days'. The origins of the phrase are interesting. The halcyon is the kingfisher of Greek legend who rather implausibly built a floating nest in the sea which heralded a fortnight of calm weather in mid December. By Shakespeare's time the phrase 'halcyon days' had simply taken on the figurative meaning of 'calm days' after a rough period. Shakespeare used the expression that way in *Henry VI Part I*, as Joan of Arc boasts that there will be sunny days to come when the English have been defeated:

> *'Assign'd am I to be the English scourge.*
> *This night the siege assuredly I'll raise:*

Expect Saint Martin's summer, halcyon days,
Since I have entered into these wars'.

So how appropriate in the halcyon days of this 'Indian Summer' that I have a very close encounter with a kingfisher. Walking back from the wilder areas of the park along the riverside park, I pause at the spot where families like to feed the ducks, on a bridge where St. John's Brook reaches the river. To my surprise a kingfisher shoots over my shoulder and perches on the rail of the bridge not twelve feet away. It is a female, identified by the orange lower part of its bill. I can see that the kingfisher has bright orange feet too. She bobs as kingfishers tend to do, then shoots away up the brook, flashing her pale electric-blue back; a brilliant view of our most brilliant bird.

Having a Bath

At the margins of the pools, and on calm days at the river's edge, birds are often seen bathing. Blackbirds, starlings and pigeons seem to be particularly fond of a daily dip. They stand in the shallows with feathers fluffed out, dipping their heads and flicking their wings to send spray over their backs. Then the bedraggled bird will fly into a bush and preen, carefully nibbling its flight feathers from the quill to the ends so that the barbs are perfectly arranged.

Bathing and preening are very important to birds, to keep their feathers in tip-top condition for optimum flight and for keeping warm, by removing dirt and any feather lice, and by applying a little oil. Most birds have an oil-secreting gland, called the preen gland (or uropygial gland) underneath their tail; this is the 'parson's nose' on a chicken or goose. The bird rubs its bill against the gland and then spreads the oil over the surface of the feathers. Oil keeps the feathers flexible and aids waterproofing but it is also thought to kill bacteria and fungi.

House sparrows are well known for taking dust baths, often after a water bath, by rolling about in dry dust or soil. The dust is thought to absorb excess preen oil and remove dry skin and parasites, but the behaviour isn't fully understood. In the park there is a bare patch of ground next to the path which is a favourite spot for the local sparrows in dry weather. Tell-tale bowl-shaped hollows show where they have been bathing.

In warm weather blackbirds and song thrushes also enjoy sun-bathing, spreading their wings and tail in a sheltered sunny spot. The sun is thought to help the preen oil to trickle through the feathers. Some ornithologists have suggested that it may also draw parasites to the surface where the bird can remove them or that the ultraviolet light in the sunlight converts chemicals in the preen oil into Vitamin D. However, one could be forgiven for thinking that they simply enjoy sunbathing, just as we do. Often a blackbird will start gently panting and half-close its eyes.

I've seen blackbirds go into an even deeper trance when they are 'anting' – bathing on an anthill and allowing the ants to run into their feathers. They will even pick up ants and rub them on their wings. Ornithologists believe the formic acid in the ants might kill feather lice. And here's an interesting notion: when sunbathing or anting, birds adopt a position remarkably similar to the heraldic emblem of the Phoenix, with wings and tail fanned and head thrown to one side. In Greek mythology, the mythical bird associated with the sun and rebirth dies in the fire but then rises from the ashes. Birds have been observed bathing in fire-ash. Could that have inspired the legend?

Bird Number Eighty

A small bird is bathing at the far side of the small shallow pool at the secluded east end of the park. It looks like a great tit as the beads of water fly about its black and white head. But no. As it stops spraying itself and sits in the water looking about, I can see it is a marsh tit. Or possibly it is a willow tit. The two species are almost identical and were only separated as different species in about 1900, partly because they have different calls. They both have smart black caps, pale cheeks, small black bibs and soft buff plumage. After bathing, this one flies up into the willows and obligingly says a loud 'pitchu-pitchu' which means it's a marsh tit. The willow tit says a nasal 'chay-chay' and has become rare in recent years.

In fact the willow tit is one of the fastest declining species in Britain and a red listed bird. The British Bird Survey that collates data on breeding birds in Britain says that willow tits declined by 77% in the thirteen years between 1994 and 2007. The marsh tit is doing a little better, but not much. The British Trust for Ornithology reports that numbers have dropped by more than 50% since the 1970s. The reasons are not clear, but it seems marsh tits need scrub rather than tree canopies, and deer have reduced woodland scrub significantly in recent years. Also an increase of blue tit and great tit numbers hasn't done the marsh tit any favours. Both species can eject the timid marsh tit from nesting holes. But here is one having a bath in the park, and it brings to eighty the total number of bird species I have managed to identify in this little haven in the centre of town.

Robin Goodfellow

Back from a holiday in France I have missed some serious floods in central England as late October produced a deluge. A week of relentless rain across Wales and the English Midlands had caused the Avon to shoot across its flood plain opposite the park and pour its swollen waters into the Severn at Tewksbury which became an island in the floods. The metal flood barriers erected along the Severn held firm but apparently it had been a close call.

By the end of the month, calm weather has returned with milky sunshine and long shadows striping the paths. The clear skies mean night-time frosts. In the copse pink and purple cyclamen are defying the morning frosts; apparently such delicate little flowers, they are clearly tougher than they look.

In late October, most of the songbirds fall silent, but the irrepressible Robin Redbreast and his companion Jenny Wren just keep on singing. Year after year, the Robin has been voted Britain's favourite bird or Britain's National Bird. It's hardly surprising when he is present in everyone's garden, hopping about cheekily underfoot when the gardener is at work, looking for worms and beetles as the ground is disturbed.

Originally called the redbreast, the robin took on its affectionate nickname in Henry VIII's time when oranges from southern Europe started to appear in Britain, and the word 'orange' was adopted for that colour. Before oranges arrived it had simply been regarded as a shade of red. The robin redbreast clearly has an orange breast. Never mind. Let's just call him Robin. The name 'Robin' seems to denote cheeky boldness. The legend of Robin Hood might not have been so appealing if he had been called Graham Hood, or Kevin Hood. Shakespeare chose to name the sprightly but unpredictable rogue, known as 'Puck', as 'Robin Goodfellow'.

But in 1820 he is still the redbreast. Keats' 'Ode to Autumn' deliberately uses the old name to denote nostalgia. The poem starts with mellow fruitfulness, but ends on a different note.

> *Hedge-crickets sing; and now with treble soft*
> *The red-breast whistles from a garden-croft;*
> *And gathering swallows twitter in the skies.*

The robin's soft whistling seems to look ahead to winter winds, as the swallows prepare to leave.

November

'There is a wind where the rose was,
Cold rain where sweet grass was,
And clouds like sheep
Stream o'er the steep
Grey skies where the lark was'.
(Walter de la Mare: 'Autumn: November')

Kestrel

Grey Days

The days are getting noticeably shorter, with a chilly breeze that sends the dark brown leaves of the oaks and the paler strands of willow swirling down to form a thick carpet in the copse. The midday sun is low and pale now, dazzling silver off the rippling river. But for most of this November the sun remains obscured behind Walter de la Mare's 'grey skies', and the weather is wet with only occasional bursts of sunshine. Two cormorants seem to have taken up residence in the park, fishing along the river, then sitting at the top of their favourite ash tree drying their wings. The open grass areas are studded with black headed gulls - as many as two hundred at a time – with a few larger common gulls among them. The river is still very high, flooding some of the paths in the park and submerging the fishing pools.

Windhover

A kestrel cruises overhead and hovers over the rough grassland, tail fanned, wings fluttering and head down, studying the ground intently. A couple of black-headed gulls immediately begin to harass it, and the kestrel wheels away. Why the gulls feel the need to see off a kestrel I'm not sure. It can't be a threat to them, surely? The kestrel feeds on mice and voles it detects with its super-powerful vision. It will take small birds if it can surprise them, but nothing larger than a lark or a meadow pipit. I suppose anything with a hooked bill is fair game for the gulls and crows, nature's bouncers.

The traditional country name for the kestrel is 'windhover' because that is what it does when hunting, facing the breeze and hovering, with its head held perfectly still as though it is nailed to the sky. How it manages to do that in blustery weather is still not really understood, but it's clear that the wings and long fanned tail are working overtime to react to the air currents and keep the bird's eyes fixed on the ground looking for any movement in the grass. The kestrel's vision is pretty remarkable.

Birds have the largest eyes relative to their size in the whole animal kingdom. The eyeball is not spherical as in mammals, but flatter, giving a wider field of view and the lens is pushed further forward giving greater magnification. Birds that are preyed upon have their eyes at the sides of their heads, giving near 360 degree vision; birds of prey have their eyes facing forward, giving binocular vision that can judge distance. I suppose that makes humans a predator species. Most birds don't use their eyelids to blink, but have a third eyelid called curiously the 'nictitating membrane' that sweeps horizontally across the eye like a windscreen wiper to keep the vision clear at all times.

The kestrel also has a very high density of receptors on its retina, and has an additional trick for finding tiny mice, shrews and voles in thick undergrowth - usually from about 20 meters in the air. It seems kestrels are able to locate the trails of voles with vision that sees in an ultraviolet light spectrum beyond our visual range. Voles lay scent-trails of urine and faeces that reflect UV light, making them visible to the kestrels, so if the hovering falcon focuses on these little roadways in the grass, surely a vole will appear soon.

Kestrel numbers in Britain rise and fall quite dramatically year on year, probably reflecting the health of the populations of voles, their staple diet. In recent years kestrel numbers have been in decline, particularly in Scotland and the north of England. But these small birds of prey can still be seen regularly, often hovering over the grass verges of our main roads. The wide verges along our motorways are long strips of uncultivated land with no humans wandering about and they are often banked, providing perfect habitat for voles. I also wonder if the constant movement and noise of the traffic makes it easier for the kestrel to hover unnoticed before dropping on to its unsuspecting prey.

Talking Trees

For a few days the clouds roll away and it becomes colder under bright blue skies. The avenue of copper beeches on the northern side of the park are a glorious blaze of colour in the sunshine, gleaming red, gold, and yellow, shimmering in the breeze, and sprinkling the path with copper coloured confetti.

Beeches and oaks don't produce the same number of beechnuts and acorns each year. These fruits and seeds are called mast. There are always a few nuts to be found under these trees, but every five or six seasons is a 'mast year' when the trees have a bumper crop and produce more fruit than they normally would. One of the main theories amongst scientists for this behaviour is called 'predator sedation'. The argument goes like this. Lots of animals, such as deer, wild boar and squirrels, as well as many birds, rely on this autumn bonanza of acorns and beechnuts to fatten up for the winter and then they are more likely to have lots of offspring the following spring. If the trees hold back their nutty bounty for a few years, there won't be so many foraging animals and birds, or 'frugifores' as they are called, coming to the forest and eating all the nuts, preventing any becoming young trees.

Then, during a mast year, the trees are laden with more fruit than the animals can possibly eat, ensuring that some seeds will be left to grow. Producing great drifts of nuts in a mast year does sap the energy of a beech tree and stops its growth for a while, but as this occurs only once every five to ten years, it's worth it for the tree to ensure the production of more saplings. And trees live at a much slower pace of life than us. If an oak can be expected to live to be 500 years old, 5 years holding back its acorns is not long in tree-time. But how do the beeches along the avenue in the park know how to synchronise their mast year?

It seems the trees can 'talk' to each other in various ways. There is a fascinating book called 'The Hidden Life of Trees' by a German forester. His name is Peter Wohlleben. Don't ask me to pronounce it. He's been studying trees for a long time and explains that in a wood or forest the trees can warn each other of an attack by caterpillars by releasing chemicals into the air, and the trees downwind will immediately start producing toxins that make their leaves distasteful to that particular caterpillar species. Not only that, the trees in a wood are interlinked underground by fine roots systems and networks of fungus that can carry messages. Perhaps that's the way they decide together, 'Let's make this a mast year and party'.

Tree Rats

Talking of nuts, the squirrels in the park are very busy burying acorns to be unearthed in the depths of winter when the squirrels fancy a snack. Researchers have established that when they emerge from their dreys on a mild winter's day, they don't dig for acorns randomly, but can remember pretty well where they cached them in the autumn.

These are not the Squirrel Nutkins of the Beatrix Potter story, with red fur, cute faces, ear tufts and very bushy tales. These are the larger and more aggressive grey squirrels, sometimes rather unkindly called 'tree rats' by naturalists because they seek out birds' nests to eat the eggs or the young birds, and boldly raid the bird-feeders in your garden. And I know from experience that you don't want them making a home in your loft, where they might nibble cables and play games of tag in the middle of the night. Beatrix Potter would have had no idea how her beloved cheeky red squirrel of the tale published in 1903, would be harassed out of most of England within a relatively short time. The invaders came from the Eastern United States.

A Victorian banker called Thomas V. Brocklehurst has a dubious claim to fame. He is the first person to be recorded having released grey squirrels into the wild in Britain – in 1876 to be precise – in Henbury Park near Macclesfield in Cheshire. After a business trip to the USA, he'd brought back a pair of greys as pets. Perhaps he realised that they weren't quite as cute as they looked, or perhaps Mrs. Brocklehurst said she wasn't having those horrible creatures in her house. Whatever the reason, he let them go. But the arrival of the greys isn't all down to the Cheshire banker.

Suddenly there was a fad for having approachable squirrels to entertain visitors in the gardens of your stately home or in urban parks. They must have been trapped and shipped to England by the crate-load. In 1890, ten squirrels imported from New Jersey were released into Woburn Park in Bedfordshire; in 1902, a hundred were released in Richmond Park in Surrey; four years later ninety-one squirrels were released into Regent's Park in London. Grey squirrels often have two litters each year and young ones can breed before they are a year old. So they can multiply quickly. And they did.

By 1930, greys were established across the south east of England, and as far north as Warwickshire. They may well have turned up in St. Nicholas Park, Warwick, shortly before the Second World War, and by the end of the war they were well distributed throughout the Midlands. By the mid-1950s greys had spread throughout most of Yorkshire, and come the mid-1980s they had crossed the Pennines into Lancashire and Cumbria and had colonised Wales and most of Cornwall.

The Decline of the Reds

The problem with the flourishing of the grey squirrel in Britain was that the native red squirrels were being driven out by their bigger and fiercer American cousins wherever they came into contact. There's little evidence of direct conflicts or fights between the two species. It seems the reds were withdrawing before the grey tide because they were unable to compete with the greys for food resources. Greys also carry more body fat and can forage more frequently in the winter, again reducing the available food for the reds. But various studies into the decline of the red squirrel point to a combination of several other reasons.

The squirrel-pox virus, which can be carried by greys but is usually fatal to reds, appears to be a significant factor in the red squirrel's decline; populations fall up to 25 times faster in areas with squirrel-pox. But it seems we humans are as much to blame as the greys. Intensive farming and the loss of large tracts of woodland in the 20th century affected the native squirrels. And not content with taking away their habitat and food sources, we killed them – in large numbers.

In her book, 'Squirrels', Jessica Holm states that in 1889 nearly 2,300 red squirrels were shot in the New Forest, Hampshire, because they were considered a pest to the timber industry. But this is nothing compared with the war on squirrels waged in Scotland. Lady Lovat of the Fraser clan championed the reintroduction of the red squirrel into the northern highlands in 1844 and the squirrels flourished. Local landowners, farmers and foresters were not amused. In 1903, the friendly-sounding 'Highland Squirrels Club' was formed, 'to counter the devastation wrought in the woods of eastern Ross-shire, Sutherland and the part of Inverness north of the Caledonian Canal'. In 1920, James Richie at the Royal Museum of Scotland wrote about the Squirrel Club, "*The results of its activities are astounding, when it is recollected that three-quarters of a century ago the squirrel was unknown in the district. During the fifteen years up to the end of 1917, 60,450 squirrels had been killed.*"

Controlling the Greys

Clearly the American grey squirrel can't take all the blame for the decline of the reds. Nonetheless, in 2015, the island of Anglesey was declared completely free of grey squirrels after an 18-year programme of culling them, despite a petition signed by 140,000 people protesting at the killings. A total of 9,597 grey squirrels were killed at a cost of just over a million pounds. That's £106.18 per squirrel. An estimated 700 red squirrels remain on the island. In the same year The Prince of Wales, then the Patron of the Red Squirrel Survival Trust, announced a cull of all greys on his Duchy of Cornwall Estate. A spokesperson for the trust said, "They cause an estimated £10 million a year in damage to trees. The greys are an invasive animal and they simply shouldn't be here". So there!

In December 2019, it became illegal to release a grey squirrel into the wild, so if an injured or baby squirrel is brought into an animal rescue centre, they'll probably have to kill it! Personally I prefer more natural solutions to ensure that our remaining populations of red squirrels are protected. The right kind of woodland management is important. Red squirrels favour woods with plenty of conifers where they can extract the small seeds from the cones. Grey squirrels tend to rely on trees with large seeds that have high calorific values. So more pines, please.

But here's a really interesting development. Until very recently, the pine marten, a rarely-seen member of the weasel family that lives in trees, was restricted to the northern reaches of Scotland and the west coast of Ireland. But recently, like its relatives the polecats and otters, it has started to reclaim some of its former range, and where this native predator has moved in, there have been declines in grey squirrel numbers allowing reds to recover. There are a number of pine marten studies going on at present, asking, among other things, why they seem to catch and eat more grey squirrels than reds. It seems that pine martens mark their territory in the trees, and the red squirrels are sensitive to this scent. They either move away to another area, or feed on high alert, looking round all the time. The greys on the other hand don't seem to react to a pine marten's scent, and feed in their usual casual manner, making them easier to surprise. I would love to think that a reintroduction of the beautiful pine marten to selected areas of woodland would help control the grey squirrel numbers.

Meanwhile in the park in the centre of town, the grey squirrels seem to have no predators to worry about, apart from the buzzards perhaps, and they seem happy to pose for pictures with tails curled over their backs as they nibble a chestnut or an acorn, delighting the children. There is even a pure white albino squirrel seen occasionally in the wooded area on the south side of the park. It's a local personality. It's picture has been in the local paper and people visit the park specially in the hope of showing it to their kids – a star among squirrels.

Water Water

After two weeks of rain the flooding in central England is getting worse. The pools in the park are flooded again and the river is surly and dark, shooting debris under the bridges.

Downstream the Severn flood defences are doing their job, but in South Yorkshire, the River Don has overflowed its banks with hundreds of people having to abandon their homes to the sludgy torrent, and many more have been flooded out in Derbyshire.

In Australia, fires are raging out of control at about 200 locations. People are driven into the sea as the flames consume their homes. Some could not escape in time and have been burned to death. The bushfire season has started early and is much worse than in previous years. The Prime Minister of Australia 'isn't convinced' about global warming and is heckled when visiting fire-ravaged communities.

In Geneva, the World Meteorological Organisation, a specialist agency of the United Nations harnessing the scientific expertise of 193 countries, issues its annual Climate Report. It makes for grim reading. Professor Sir David King, Britain's former Chief Scientist, says he is 'scared' by the results. The report says the concentration of climate-heating greenhouse gases in the atmosphere has hit a record high. So far, international action on the climate emergency is having no effect in the atmosphere. The WMO says despite all the commitments made under The Paris Agreement of 2015, the gap between targets and reality were both 'glaring and growing'. Global emissions continue to 'surge'. The world's scientists calculate that emissions must fall by half by 2030 to give a good chance of limiting global heating to 1.5°C, beyond which 'hundreds of millions of people will suffer more heatwaves, droughts, floods and poverty'. Well, halving emissions in a few years is just not going to happen. It will be the poorest countries that suffer the most from global warming.

I feel very fortunate to live where I do. The Avon soon returns to where it ought to be, the dark grey clouds move away, and by the end of the month, the days dawn crisp and bright with blue skies.

Water Everywhere

Early in the morning the river is on fire, reflecting the sunrise with glittering orange streaks below clouds of mist smoking off the still surface. Autumn mist is always lovely to see, and is regarded as rather Romantic in the poetic sense. One half-expects to see the pale swirling clouds part to reveal the Lady of the Lake at the prow of a barge, holding out Excalibur, or perhaps in Warwick it should be Aethelflaed, the Anglo-Saxon Lady of the Mercians who founded the town.

But what causes the river and the pools to 'smoke' in this way on a crisp November morning? It's quite a complex phenomenon. On a cold night, the water is relatively warm compared with the air. Water is constantly evaporating, releasing tiny beads of water molecules in gas form into the air. But cold air cannot hold as much water as warm air, so as the air reaches 100% humidity, these minuscule water droplets condense in the cold air and form mist. As the rising sun warms the atmosphere, the air is capable of holding more water vapour in its invisible form, and the mist fades away.

Water is, of course, the main reason there is life on Earth. In its many forms water is everywhere. The surface of the planet is 71% covered in water; that is why, from space, it is The Blue Planet. An astonishing 96.5% of all the water is saltwater, contained in our oceans. The remaining 3.5% is in the air as water vapour, in our freshwater rivers and lakes; it is held in our icecaps and glaciers, and is in the ground as soil moisture. It is also in us and the rest of the animal kingdom.

On average, the human body is 60% water. Small children are 75% water, probably because they contain less fat, gristle and bone, though bone has water in it too. The British Isles are watery places. This green and pleasant land is surrounded by salt water, and is constantly washed with fresh water as the weather systems arrive on the jet stream. I reckon we should all be thankful for the amazing abundance of wildlife in all parts of the country, including the very heart of England, the cradle of the industrial revolution that changed the world.

December

'Come, come thou bleak December wind,
And blow the dry leaves from the tree'.
(Samuel Taylor Coleridge).

Robin

Eerily Quiet

It is still cold, clear and calm with the ground saturated, (it was the fifth wettest November since records began), but the river is within its banks and it has now stopped. How can that happen, with flotsam including large branches of trees not moving at all? The river has become torpid, in stasis, too lazy to make its way to the sea, or probably the swollen Severn downstream is causing a bit of a water logjam, backing up its tributaries. The birds are in a semi-state too. It is eerily quiet, apart from the occasional piercing whistle of the redwings in the treetops and one or two twittering robins.

Middle Earth

A loose flock of 9 cormorants flap their way high over the river, reminding me of the hideous flying steeds of the Nazgûl, the nine Ringwraiths or Black Riders who served the evil Lord Sauron. Alright, I know some people don't like 'The Lord of the Rings', but as soon as I picked up the books as a schoolboy, I loved the world created by J.R.R.Tolkien based on Nordic myths; and of course The Shire inhabited by the bucolic Hobbits is right here – an idealised version of the English Midlands he knew well, having been brought up in south Birmingham.

Tolkien married his teenage sweetheart Edith Bratt at Mary Immaculate Roman Catholic Church in Warwick in March 1916, and three months later he was on his way to France and the Battle of the Somme. At the end of the four-month carnage, more than a million British, French and German soldiers were dead. Tolkien returned to England suffering from trench fever, a debilitating disease transmitted by body lice. He spent the rest of the war alternating between hospital and light garrison duties before becoming a university professor and a leading authority on Anglo-Saxon and Middle English language and literature.

The war memorial in the centre of Warwick carries 365 names of local men, and one woman, who died serving in WW1, many on the Somme in 1916. The Royal Warwickshire Regiment, one of the biggest infantry regiments, (Birmingham was a part of Warwickshire in those days), fought on the Western Front throughout the war, and lost a staggering total of 11,610 men. A few years later, the small town of Warwick lost another 112 men in WW2, some flying dangerous bombing missions over Germany, some in North Africa and the Middle East, some in Italy, and some, once again, on the Western Front in France. Standing by the Kingfisher Pools on a calm morning, it is easy to reflect on the sacrifices made by my grandparents' and parents' generations to protect Britain's way of life in two world wars, and the stupidities and cruelties that caused them.

The Elusive Bullfinch

In the quiet of a windless morning I can hear a soft 'phu' single note call in the thicket, answered by another further down the path. This is unmistakably a pair of bullfinches keeping in touch with each other. For such brightly-coloured birds they are difficult to see, keeping in cover most of the time and wary of people. But there in the hawthorn I can glimpse the male, with his bright pink chest, grey back and black crown. The female has the same black crown but is grey and buff; both have bright white rumps and pale wing bars visible when they fly off. The name comes from the 'bull head'. This is a bird with no visible neck and a stubby bill. A pair of bullfinches becomes strongly attached. If you see one, you are very likely to see or hear its partner.

The bullfinch has a pretty song when it is in the mood, though it is not as vocal as the goldfinch, and apparently it can be taught to mimic other sounds, so it became a popular cage bird in Victorian times. In Britain the population of bullfinches fell by 35% in the 50 years from the early 1970s possibly due to intensive farming practices, but this attractive bird has recently showed some signs of the green shoots of recovery. I mention green shoots because this slight increase in numbers is not exactly welcome in some quarters; the bullfinch likes to eat green buds and is no friend of fruit-growers.

Downstream from Warwick and Stratford, the Avon meanders through a fertile valley called The Vale of Evesham, famous over many centuries for its wide variety of soft fruit – plums, strawberries, cherries, gooseberries, pears and apples, produced in huge quantities. Fruit growing here in the sheltered vale with its rich soil began with the Benedictine monks of Evesham Abbey in the eighth century. Surviving the dissolution of the monasteries and the ravages of the Civil War, fruit farming had a boost first from the industrial revolution that brought big population increases in nearby Birmingham and the Black Country, and secondly from the arrival of the Cotswold Railway 150 years ago, making delivery of fresh fruit and vegetables to the big cities easier and more profitable. At its height, the whole of the Vale of Evesham was covered in blossom in springtime and became a tourist destination. Unfortunately for the bullfinches, they too were attracted by the spring blossom, not to admire the view but to feast on the buds. Growers say a pair of bullfinches can strip an apple tree in a few hours.

To this day, 'Natural England' issues licences for culling bullfinches on a case-by-case basis, where there is evidence of major damage to fruit trees. They can be shot, or trapped and then humanely destroyed. Fortunately there are many less lethal methods for keeping the finches away; bird scarers seem to have some effect. But after the bang that sends the pigeons clattering away, the small birds seem to return after a short while to keep feeding on those tasty buds. Netting or 'cottoning' the trees and bushes is labour-intensive and the nets or networks of cotton-strands can entangle birds. But there is a cheaper and more effective deterrent. Farmers say hanging strips of silver paper from the branches does the trick, and best of all are plastic bags attached to the branches. They bounce about in the breeze like beach balls and making a fluttering noise, unnerving the birds. So if you drive through the picturesque Vale of Evesham when the blossom is appearing, don't be surprised to see the orchards festooned with rubbish. It's all part of the cunning plan to give you fresh fruit in the supermarket and apple cider in the farm shops.

Peregrines on the Tower

Mid-December and there is no sign yet of any very cold winter weather, let alone snow. It is calm with milky sunshine and some light showers. Walking towards the park I hear the mewing 'kea-kea' call of a peregrine above, interspersed with angry whistling sounds, and there are two birds heading away from St. Mary's Church tower. They are the same compact size, so must be two males. One bird is repeatedly swooping at the other; the second falcon rolls over to show its talons as the noisy assailant shoots past. They head away over Priory Park with the aerial battle continuing. It seems fairly clear that the angry bird was defending its territory – the church tower. Could this mean it sees it as a potential nest site for the spring? That would be exciting.

A few days later, on a still, grey afternoon, two peregrines are circling the St. Mary's tower. One is noticeably larger than the other – a female. She perches on a ledge. The male makes a number of close passes, some at speed, and perches briefly on various places on the tower, including one of the pinnacles. Is this a pair prospecting for a nest site? Or do they just regard this as their winter roost? I should know not long after Christmas. The birds will have decided on their nest site by the end of February, and the female will start laying her eggs in late March.

The Christmas Robin

In all corners of the park, the robins are singing their shrill chirruping song. The nation's favourite bird is feisty and bold, and Cock Robin's cheerful piping notes can be heard all year long, even in the middle of winter, and even in the middle of the night. In fact the song is a warning to other males to keep away as he stakes his claim. The European robin does not migrate, and when a male has established a territory, he will defend it fiercely, chasing potential rivals with his angry 'tic-tic' warning call, posturing to show off his red breast, and if that doesn't work, engaging in full-on combat. There are many observations of robins fighting to the death. They have even injured themselves trying to fight their reflections in car wing-mirrors. But as the season of peace and goodwill approaches, we see the robin as a loveable friend dressed in festive red, with his perky image adorning millions of Christmas cards.

The tradition of robins on greeting cards goes back to Victorian times when the postmen wore a red uniform, and were nicknamed 'robin redbreast the postman', As Christmas drew near, families eagerly awaited cards from their loved ones, and looked out for their local 'robin'. Artists began illustrating the cards with robins delivering cards and letters, and they soon became a Christmas icon. But the robin's association with

Christmas goes back a lot further. Christian folklore has it that the robin was a brown bird who, distressed to see Christ's suffering on the cross, tried to remove the crown of thorns while singing a comforting tune and became bloodstained – a similar legend to that of the goldfinch with its bright red face.

The Midwinter Festival

Many Christian stories have adopted pre-Christian symbols and celebrations. The robin – singing brightly during the shortest days – might well have been a symbol of the turn of the year in times unrecorded. The middle of winter has long been a time of celebration around the world. The winter solstice, when the days start to get longer, is a time of feasting in many cultures. In Scandinavia, the Norse celebrated 'Yule' from the solstice through January. Fathers and sons would bring home large logs to be thrown on the fire, and families would get together to feast until the log burned out – which could be several days – hence the yule log. It was a good time for a feast. Pigs and cattle were slaughtered so that they would not have to be fed and tended during the bad weather, and to provide a larder for the worst of the winter months. And handily, most of the wine and beer made during the year was fermented and ready for quaffing.

In ancient Rome at this time they celebrated, 'Saturnalia', a holiday in honour of Saturn, the god of agriculture. Continuing for a full month, the normal social order was turned upside down, with slaves becoming masters and peasants in command of the city. And in particular Romans celebrated December 25th, the birthday of Mithra, the god of the unconquerable sun. Is this why Pope Julius I sitting in Rome in the 4th century chose December 25th as the birth date of Christ? The Bible does not mention a date. By holding Christmas at the same time as the traditional winter solstice festivals, church leaders increased the chances that Christmas would be embraced and celebrated. And it certainly was. Now about 2 billion people across the world celebrate Christmas on December 25th, many of them in the UK by eating a bird, and it's almost guaranteed that the cards hanging from the mantlepiece, or in Australia the air-conditioning unit, will include at least one picture of a robin.

An Exciting End to the Wildlife Year

The traditional Boxing Day walk has been a washout with relentless rain, and the river is lapping over its banks again, but December 28th is dry and calm, so I venture out to walk off some of the yuletide excess, and to see what's happening in the park. The Cetti's warbler is wasting no time in belting out its song to attract a mate, and a few snowdrops are starting to show already. It looks like it could be another winter with no snow.

Walking along the path by the river, I noticed about twenty black-headed gulls and two magpies screaming and chattering at something on the other side of the river. It was somewhere in the bank under some leaning willows. Could it be a cat? Or more likely a rat? Then a sleek, dark shape slipped out of the water on to the bank – an adult otter! I could see through my binoculars that it had a small fish that it proceeded to eat showing its sharp teeth and pink tongue, ignoring the noisy attention of the gulls. Throughout my life I had seen otters only in Scotland, and rarely – four distant views at most. It was thrilling to see this elusive wild animal at quite close range so close to home. The otter slid back into the water and worked its way along the far bank downstream towards the castle, past the landing stages, then turned back and disappeared.

My morning walks in the park had a new focus. Was this a one-off sighting, or had an otter made this reach of the river part of its territory? A few mornings later, there it was again, this time working its way towards me along my side of the river. I stood still on the concrete bank waiting for it to reach me, and sure enough it popped its head out just four feet away. I could see every detail of its long whiskers and beady eyes. It immediately dived and crossed to the far side of the river. I grabbed a bit of indifferent video on my phone's camera and the otter disappeared. It seems to have a habit of diving and disappearing.

In early January there were more astonishingly close views of this very shy animal. One grey morning I had given up looking for the otter on my morning walk and was starting back along the river path when the magpies and crows began chattering and cawing upstream, attracting some black-headed gulls. I hurried back to the footbridge, and sure enough, the otter was working its way downstream towards me along the left bank of the river. It paused immediately below me on the roots of an alder, then swam under the footbridge and up the centre of the river followed by some gulls. I walked quickly past the old Sea Scout hut and along the park path where I met my friend Alan from the Natural History Society. He was envious about my otter sightings and had been out each morning looking for it, without success. I pointed to a dot in the middle of the river and he was delighted. We could see the otter diving in mid-stream next to the copse. We hurried there and watched it swimming into our bank just below us in the tangle of roots. The otter looked directly at us several times but seemed undisturbed and groomed itself for several minutes before slipping back into the water and disappearing.

Alan said, "I never would have believed I would see a wild otter ten minutes' walk from my front door". I feel the same. And it's such a cool animal, if you know what I mean – sleek, casual, confident and seems to know it is good looking. I sent the sightings to the Warwickshire Biological Records Centre. They have local otter sightings dating back to 2004, but they have mainly been in the River Leam at Leamington with nothing recorded in the centre of Warwick. It seems this otter is the first to be seen in the centre of Warwickshire's historic county town for many a long year.

An Extraordinary New Year

'What's important is to be prepared'.
(Hamlet, Act V Scene 2)

Collared Dove

The Invisible Killer

While I was happily watching a wild otter in the park in Warwick on New Year's Eve, I could not have known that on that same day the authorities in China were alerting the World Health Organisation to several cases of unusual pneumonia in a place called Wuhan, a city of 11 million people not well known in the west. Several of those affected worked in the city's Seafood Wholesale Market, which was shut down the following day. This information did not feature on the main news in Britain as the spectacular firework displays around the globe heralded a new decade. We could have no idea that this unknown virus would change the world.

Covid -19 was to prove more deadly than the Sars outbreak of 2002-3 which claimed 774 victims worldwide. The eventual death toll from Covid-19 was likely to be counted in hundreds of thousands, perhaps even millions. In Britain, after a complacent and disorganised response, the nation was told to 'Stay at Home' and an unprecedented lockdown began, separating families and friends and crippling the economy. The streets became eerily silent and empty. There was no traffic noise. There were no planes roaring across the sky; no vapour trails in the blue. The number of coronavirus deaths rose relentlessly, many of them in care homes which had been eclipsed by the authorities' priority of protecting the NHS.

Life was on hold in what has been called 'the great pause'. But Nature was not on hold. This all happened in springtime, and suddenly the newspapers were full of letters saying readers had never really appreciated the birdsong before. Now they could hear it. And they had plenty of time to stop, listen and look. The press responded by publishing articles on how to identify your garden birds, with some samples on their websites. After a few weeks of lockdown, the problem of mental health became prominent. A high proportion of people reported anxiety, loneliness and depression. Children separated from their friends and grandparents and deprived of exercise were a particular concern. The answer, said the experts, was access to greenspace and regular walks in a natural setting. It seems it needed a global disaster to reconnect people with nature for their physical and mental wellbeing.

The haze of pollution in many of the world's cities from New Delhi to Beijing, and Paris to Madrid, disappeared and there was clear air for the first time in decades. According to the Centre for Research on Energy and Clean Air, the first month of lockdown saw a dramatic reduction in major air pollutants across Europe compared with the same period the previous year. Portugal led the way with a 50% drop in nitrogen dioxide in the air. There were fewer cases of asthma, particularly in children, and fewer cases of heart and respiratory problems caused by particulates in the air lodging in the lungs. The statisticians estimated that the cleaner air across Europe led to 11,000 fewer deaths – not many compared with the coronavirus numbers, but that was only for one month of less polluted air.

A Different Future?

The pandemic forced countries around the world to enact strict lockdowns, seal borders and scale back economic activities. Analysts soon found that these measures contributed to an estimated 17% decline in the daily global carbon dioxide emissions that cause climate change, compared with the daily global averages the previous year. It's a worldwide drop that scientists say could be the largest in recorded history.

The world is struggling to cope with the impact of the pandemic. Surely nations should try to establish a different world order rather than return to the economic models of the past that were based on more and more consumption and mountains of waste? Will this disaster be a spur to develop new industries and new jobs supporting healthier lifestyles? Could it accelerate the move to greater energy-efficiency and a circular economy that uses waste as a resource?

With the world economy paralysed, the billionaire Chief Executive of J.P.Morgan, America's biggest bank, rather unexpectedly calls the pandemic "a wake-up call ... for business and government to think, act and invest for the common good". Has it shown a way to reduce stressful and polluting commuting, with more people working from home; will there be more investment in cycling and better public transport?

The main question for the developed world is whether people will change their behaviour. Will the shock of Covid-19 make us more aware that – when we have to confront mortality – we value people more than things, and that we value most those who contribute most to our societies?

And what about Climate Change? Will there be a new impetus behind switching to all-electric vehicles based on renewable energy? After all, the huge threats posed by global warming continue to hang over the next generations, and could ultimately cause much more social damage than Covid-19. After the inevitable worldwide recession, can governments afford to make the changes? Can they afford to invest in a green economy that really does reduce greenhouse gas emissions; or perhaps the question should be, will they choose to?

In the English Midlands there was to be no snow at all during the rest of the winter. Scientists reported that the year just past had been globally the second hottest on record. After mid-January the otter was not seen again. The pair of peregrines was back on the church tower using it as a winter roost and feeding place, but again, as spring arrived, they were to nest elsewhere. Soon the birds were singing once more as the yearly cycle of renewal began and the park would become full of life and colour once more.

The pandemic had shown beyond doubt the benefits of cleaner air and access to 'greenspace'. Let us hope the world recovery respects our natural environment on Earth, with the focus on improving people's wellbeing and health as well as wealth. It will not be 'a walk in the park', but for me, the daily contact with my local patch of nature certainly helps me to cope with the unfolding world events. Wild spaces in towns and cities have never been more important.

The Author

Rick Thompson is a former journalist and broadcaster and a lifelong birdwatcher. He worked for BBC News for 27 years, starting at BBC Birmingham where he produced and presented regional wildlife programmes as well as reporting the news. He moved to London where he became a senior editor with BBC Television News and The World Service. Later he returned to the Midlands as the BBC's Head of News, Current Affairs and Local Programmes.

For several years Rick wrote and illustrated a regular column on birds for the Countryside magazine, and for four years he was a member of the Governing Council of the RSPB. He is regularly asked to moderate European conferences on the environment, sustainable energy, transport and climate change.

He says, "I believe that exposure to nature is incredibly important for people's physical and mental well-being, and getting to know the birds and other wildlife near your home is fascinating and uplifting. With four out of five people in Britain living in towns and cities, urban green spaces are vital for our health and happiness. I hope this diary of a year in a large town park will encourage people to get out into the natural environment, and enjoy the changing seasons and the marvellous variety of creatures and plant life that can be found close to home. I also hope that planners, developers, farmers and local authorities will protect our valuable wildlife habitats and invest in more green spaces that are accessible to all".